Pus

Push

*Software Design and the Cultural Politics
of Music Production*

MIKE D'ERRICO

Oxford University Press is a department of the University of Oxford. It furthers
the University's objective of excellence in research, scholarship, and education
by publishing worldwide. Oxford is a registered trade mark of Oxford University
Press in the UK and certain other countries.

Published in the United States of America by Oxford University Press
198 Madison Avenue, New York, NY 10016, United States of America.

Library of Congress Cataloging-in-Publication Data
Names: D'Errico, Mike, author.
Title: Push : software design and the cultural politics of music production / Mike D'Errico.
Description: [1.] | New York : Oxford University Press, 2022. |
Includes bibliographical references and index.
Identifiers: LCCN 2021033940 (print) | LCCN 2021033941 (ebook) |
ISBN 9780190943318 (paperback) | ISBN 9780190943301 (hardback) |
ISBN 9780190943332 (epub) | ISBN 9780190943356 | ISBN 9780190943349
Subjects: LCSH: Popular music—Philosophy and aesthetics. | Electronica
(Music)—Philosophy and aesthetics. | Popular music—Production and
direction. | Electronica (Music)—Production and direction. |
Music—Computer programs. | Music and technology.
Classification: LCC ML3877 .D27 2022 (print) | LCC ML3877 (ebook) |
DDC 781.64/117—dc23
LC record available at https://lccn.loc.gov/2021033940
LC ebook record available at https://lccn.loc.gov/2021033941

DOI: 10.1093/oso/9780190943301.001.0001

To Bess, Lily, and Joanie, for pushing me.
To Gary Michael D'Errico, in loving memory.

Contents

Acknowledgments

First and foremost, thanks to my wife, Bess, for her endless patience and dedication as I pursued the uncertain path of graduate school and the early stages of an academic career. Between the time I began work on this book and the time I submitted the final manuscript, we moved across the country twice, got married, and had two kids. All the while, Bess worked multiple jobs and watched the kids to make sure I could attend conferences, travel to job interviews, and pursue the research that ultimately led to this book. This is the invisible labor that supports so much academic work, and nobody knows how to hustle like Bess. To our daughters, Lily and Joanie: this book represents one snapshot in a long journey, the high points of which were the births of you two. You made it all worth it, and you provide the spark that carries us through.

To my mom, Becky Suchovsky, and my dad, Scott "TooTall" Suchovsky. To my brother, Chris D'Errico, and my sister-in-law, Tamila D'Errico. To my mother-in-law, Martha Brooks, and my father-in-law, Richard Brooks. To my sister-in-law Kate Neilsen, my brother-in-law Joey Neilsen, and my two nephews, Owen and Levi. To the Berubes, D'Erricos, Dicksons, Godinezes, Hanlons, Henshaws, Hornes, Jabours, LaBelles, Parottas, Salinders, Thorntons, Urbaneks, and Zukovics. Thank you all for your endless support, even when the end goal wasn't always clear.

To the teachers who got me into this in the first place: Andy Boysen, Ken Clark, Mark DeTurk, Tony DiBartolomeo, Lori Dobbins, Robert Eshbach, Bruce Gatchell, Les Harris Jr., John Herman, Bill Kempster, Tim Miles, David Newsam, Matt Redmond, David Ripley, Andy Robbins, Dave Seiler, Mark Shilansky, Nancy Smith, Mike Walsh, Richard Young. Thank you for continuing to inspire me.

To the academic mentors who pushed me along the way: Joseph Auner, Dan Beller-McKenna, Jane Bernstein, Olivia Bloechl, Alessandra Campana, Johanna Drucker, Nina Eidsheim, Robert Fink, Elisabeth Le Guin, Rob Haskins, Mark Katz, Raymond Knapp, Tamara Levitz, Peter Lunenfeld, David MacFadyen, Stephen Mamber, Annie Maxfield, Kiri Miller, Mitchell Morris, Stephan Pennington, Miriam Posner, Todd Presner, Jessica Schwartz,

Jonathan Sterne, Jeffrey Summit, Tim Taylor, Elizabeth Upton. Thank you for the wisdom and for sparking in me new layers of curiosity.

To my colleagues in the Albright College Music Department: Dana Allaband, Tammy Black, Tim Gross, Jeff Lentz, Mark Lomanno, AJ Merlino, Kelly Powers, Jordan Shomper. To my colleagues from the UCLA Music Industry program: Jonathan Beard, Gigi Johnson, David Leaf, Adam Moseley; and from the Intercollegiate Media Studies program at the Claremont Colleges: Elizabeth Affuso, Mark Andrejevic, Eddie Gonzalez, Stephanie Hutin, Ian Ingram, Ann Kaneko, Jesse Lerner, Ming-Yuen S. Ma, Nancy Macko, Ruti Talmor, T. Kim-Trang Tran, Carlin Wing; and from the Pomona College Music Department: Alfred Cramer, Thomas Flaherty, Genevieve Lee, Joti Rockwell, Gibb Schreffler. Thank you for the opportunity to build together.

For the conversations in graduate school, from Davis Square to Echo Park, to Francesca Albrezzi, Alexandra Apolloni, Patrick Bonczyk, Caitlin Carlos, Hyun Kyong Chang, Monica Chieffo, Marci Cohen, Benjamin Court, Patrick Craven, Roderic Crooks, Wade Dean, Andrew deWaard, Albert Diaz, Benjamin Doleac, Megan Driscoll, Oded Erez, Rose Fonkham, Rebekah Lobosco Gilli, Ian Goldstein, Gillian Gower, Alex Grabarchuk, Pheaross Graham, Stephanie Gunst, Tom Hanslowe, Brendan Higgins, William Hutson, Jake Johnson, Pradeep Kannan, Kyle Kaplan, William Kenlon, Ben Krakauer, Wendy Kurtz, Melinda Latour, Scott Linford, Joanna Love, Kimberly Mack, Alyssa Matthias, Dave Molk, Andrea Moore, Kacie Morgan, Warrick Moses, Tiffany Naiman, Marissa Ochsner, Julius Reder-Carlson, Terri Richter, Alex Rodriguez, Jill Rogers, Mehrenegar Rostami, Anahit Rostomyan, Chris Santiago, Eric Schmidt, Darci Sprengel, Danielle Stein, Otto Stuparitz, Christiana Usenza, Kristie Valdéz-Guillen, Jessie Vallejo, Zachary Wallmark, Schuyler Dunlap Whelden, Dave Wilson, Morgan Woolsey. Thank you for the invaluable insights, perspectives, and experiences.

Thanks to the academic friends I've made along the way: Tim Anderson, Christine Bacareza Balance, Eliot Bates, Samantha Bennett, Christa Bentley, Matt Brennan, Erin Brooks, Matt Brounley, Ragnhild Brøvig-Hanssen, Justin Burton, Joshua Busman, Mark Butler, Daphne Carr, Theo Cateforis, Jace Clayton, Norma Coates, Amy Coddington, Ali Colleen, James Currie, Anne Danielsen, James Deaville, Zach Diaz, Jess Dilday, Emily Dolan, Kira Dralle, Jarek Ervin, Mike Exarchos, Rebekah Farrugia, Jessica Feldman, Andy Flory, Murray Forman, Kate Galloway, Luis-Manuel

Garcia, Rolf Inge Godøy, K. E. Goldschmitt, Wells Gordon, Sarah Hankins, Anthony Kwame Harrison, Jasmine Henry, Madison Heying, Jessica Holmes, Umi Fangyu Hsu, Robin James, Brian Jones, Katherine Kaiser, Loren Kajikawa, Sarah Kessler, Eve Klein, Adam Krims, Elizabeth Lindau, Alejandro Madrid, Andrew Mall, Noriko Manabe, Caitlin Marshall, Wayne Marshall, Katherine Meizel, Esther Morgan-Ellis, Darren Mueller, Kjell Andreas Oddekalv, Justin Patch, Sean Peterson, Benjamin Piekut, Catherine Provenzano, Guthrie Ramsey, Chris Reali, Katie Reed, Anders Reuter, Griff Rollefson, Bjørnar Sandvik, Margaret Schedel, Justin Schell, Martin Scherzinger, Joel Schwindt, Victoria Simon, Mary Ann Smart, Jason Stanyek, Victor Szabo, Nick Tochka, Jon Turner, David VanderHamm, Lisa Cooper Vest, Steve Waksman, Elijah Wald, Oliver Wang, Eric Weisbard, Christy Jay Wells, Justin Williams, Griffin Woodworth, Brian Wright, Simon Zagorski-Thomas.

Thanks to the friends, artists, and musicians who continue to inspire me: Roland Adams, Yvonne Aubert, Ryan Baker, Yeuda Ben-Atar, Anders Benson, Patrick Boutwell, Jon Briggs, Chris Burbank, Bobby Burns, Chris Chase, James Clark, Dennis Coffey, Andrew Cote, Chris Coté, James Dattolo, Stuart Dias, David Dodson, Jim Dozet, Stephen Dunleavy, Mike Effenberger, David Evans, Brett Gallo, Margot Garber, Mary Gatchell, Rob Gerry, Antonio Giamberardino, Nick Gilbert, Adam Gouveia, Andy Grant, Andy Greene, Sharon Harvey, Eben Hearn, Chris Hislop, Fred Horne, Juliet Hulme, Sarah Strosahl Kagi, Daniel Katsekas, Ben Keeler, Chris Klaxton, Eric Klaxton, Frank Kumor, Jesse Lannoo, William Lawrence, Jon and Amanda Ludwig, Nick Mainella, Adam Marcello, Kendall Martin, Jake Mehrmann, John Mehrmann, Chris Milner, Jen Mitlas, Darryl Montgomery-Hell, Danielle Moreau, Caitlin and Aaron Morneau, Wayne Moulton, Daniel Murphy, Jeremy Murphy, Joseph K. Murphy, Roland Nicol, Eric Von Oeyen, Carl Peczynski, Nick Phaneuf, James and Lydia Primate, Brittany Fecteau Robert, Frank Roberts, Matt Salinder, Dedrick Shimke, Chris Sink, Jonathan Snipes, Jon Starr, Didier Michel Sylvain, Jason Trikakis, Nick Trikakis, Zach Wilson, Chris Yesno.

Special thanks to the industry friends and partners who were willing to share ideas with me: Alfred Darlington, Vincent Diamante, Richard Flanagan, Claire Lim, Kevin Marques Moo, Miller Puckette, Rich Vreeland, Josh Weatherspoon; the folks at Ableton: Javad Butah, Ben Casey, Michael Greig, Tony McCall, Matthew West; and the folks at Avid: Bryan Castle, Frank Cook, Andrew Hagerman, Eric Keunhl, Shilpa Patel, Daniella Singh.

Thanks for the collaborative research to my students Jonathan Craft and Sarajean Reinert and to the Albright Undergraduate Research Council for funding projects that led to many insights presented in chapters 2 and 6. Thanks to the Albright College Professional Council for development funds that helped support the publication of this book.

Finally, thanks to my editor, Norman Hirschy, whose unshaken faith in this project encouraged me to write the book before I knew what it would become.

Portions of *Push* were first published in the following publications:

"Diggin' in the Carts': Technologies of Play in Hip-Hop Production and Performance." In *Critical Approaches to the Production of Music and Sound*, edited by Samantha Bennett and Eliot Bates. London: Bloomsbury, 2018. Reprinted with the permission of Bloomsbury Academic.

"Worlds of Sound: Indie Games, Proceduralism, and the Aesthetics of Emergence." *Music, Sound, and the Moving Image* 9, no. 2 (January 2016). Reprinted with the permission of Liverpool University Press.

About the Companion Website

www.oup.com/us/push

Oxford University Press has created a website to accompany *Push*, which includes video demos and overviews of the software and hardware described throughout the book. Multimedia examples available online are found throughout the text and are signaled with ⊙.

Introduction

Interface Aesthetics

In the 2009 report, "90% of Waking Hours Spent Staring at Glowing Rectangles," satirical news outlet the Onion provides an account of the ubiquity of hardware and software devices in the everyday lives of Americans. Most noticeably, the author highlights the pervasiveness of technological objects throughout a variety of daily activities, describing "handheld rectangles, music-playing rectangles, mobile communication rectangles, personal work rectangles, and bright alarm cubes." [1] While the piece may serve simply as comic relief for a wired generation, all too familiar with navigating the daily grind through the mobility of laptops and smartphones, for others it pinpoints many of the anxieties surrounding an emerging digital maximalist lifestyle. Whether one embraces or resists the contemporary pervasiveness of digital media, the piece exposes a fundamental tension in our interactions with the digital world: as material technologies increasingly move to the screen space of software, what creative strategies can be employed to maneuver interconnected and complex modes of practice? In short, how does one engage with media after software?

For many digital artists, technological design and practice in the twenty-first century are defined by the convergence and cross-pollination of media, resulting in the simultaneous centralization of tools and techniques on laptops and other digital workstations and the proliferation of interfaces and peripherals for physically engaging that software. Music producers' bedrooms are semiprofessional studios, just as mobile smartphones offer full-fledged audio recording capabilities; DJs are increasingly expanding the size of their performance setups while learning how to spin records from "the cloud" with a tablet computer; a two-thousand-dollar vintage synthesizer is now available as a ten-dollar "app." In response to this shift from hardware to software, artists, musicians, and technologists of various sorts have developed new interfaces for interactively engaging and embodying digital materials, from touchscreen devices such as smart media to programming

Push. Mike D'Errico, Oxford University Press. © Oxford University Press 2022.
DOI: 10.1093/oso/9780190943301.003.0001

languages for the creation of digital art such as *Max* and *Processing*. As sound continues to become enmeshed in the ever-changing screen space of software, how have musicians and software designers adapted existing modes of engagement to musical interfaces? Moreover, how has the emergence of multimodal interfaces introduced new conceptions of what constitutes "sonic" material in the digital age? At a fundamental level, we might ask, what is sound after software?

It was this question that brought me to Los Angeles in 2011. At the time, the city was experiencing a renaissance in two seemingly unrelated cultures, both as a result of developments in media production software: first, hip-hop and electronic dance music (EDM) and, second, game design. With the passing of beatmaking pioneer J Dilla in 2008, the East Los Angeles experimental nightclub Low End Theory became a hotbed for musicians, video artists, and hip-hop producers. Located far outside the glitz and glamor of the Hollywood EDM scene, Low End Theory was always on the cultural fringe. Most significantly, the club focused on instrumental hip-hop and dance music performance, showcasing the technological prowess of beatmakers rather than the lyrical poetics of rappers. Indeed, the venue originated as an artists' collective dedicated to the *process* of creating art, and it mainly attracted "beatheads" more interested in learning techniques from the artists onstage than in dancing to their music.

While hip-hop practitioners had always admired the perceived physicality of "analog" hardware devices such as drum machines and samplers, the experimental nature of Low End Theory led many DJs to embrace software in their stage setups for the first time. As a result, the laptop screen created a literal and metaphorical barrier between performer and audience, encouraging musicians to seek new ways of highlighting the process of music making for their audiences. Countless debates ensued concerning the "proper" ways in which to perform hip-hop and dance music, with professional musicians and fans alike constructing rhetorical dichotomies between the perceived "authenticity" of analog hardware versus the ephemerality of digital software. Like the introduction of the turntable to hip-hop thirty years earlier, the emergence of software in DJ culture marked yet another moment in which a new technology negotiated its changing status as a musical instrument.

Meanwhile, a group of students at the University of Southern California were radically reinventing gaming aesthetics. Small-scale releases from the independent game development team Thatgamecompany consciously moved away from the adrenaline-fueled, hypermasculine ethos of first-person shooters toward a nonlinear experience in which players explore the virtual world on their own terms. Games such as *Flower* (2009) and *Journey* (2012) foregrounded unique gameplay mechanics over narrative, highlighting the process-oriented nature of "indie" video games in the new millennium. Players learned the rules of these games through indirect exploration rather than following strictly predetermined cinematic storylines. Sound design for the games followed suit, aurally guiding the player through the virtual world in the absence of direct visual cues. As the design of the games mimicked the algorithmic nature of software, sound became more than just background noise; it became a gameplay mechanic in itself, a sonic manifestation of the rules embodied in the "hardware" of both game console and player.

For most of the 2010s, I studied these two cultures through interviews with beatmakers, game developers, and media artists, as well as through participant observation as a sound artist, DJ, and game scholar myself. I was well positioned to understand the aesthetics of these communities. In the decade before moving to Los Angeles, I had produced beats for rappers, performed experimental multimedia sets, and composed video-game music using vintage game consoles. But I only learned so much by focusing on technologies of production from the perspective of a producer. By examining the Los Angeles music and gaming cultures from the perspective of a scholar-practitioner, I realized a few things about the nature of human-computer interaction (HCI). First, software is more than just a tool. It's a platform for the expression of individual and collective identities, values, and politics. Second, the relationship between old and new media is based not simply on *remediation*, in which the new refashions the old, but also on *accumulation*, as tools and techniques continue to converge in software. Most significantly, I began to understand the interfaces that surrounded myself and my work as more than just control surfaces; they were pivotal components of a saturated media environment that was increasingly altering the practices of artists and media professionals. These insights led to this book.

Push examines the aesthetics and technical practices of digital audio producers from multiple, sometimes conflicting perspectives, including musicians, music scholars, sound artists, software developers, and game designers, as well as my own experience working with sound across media. The term "push" carries multiple meanings. For those in the know, it likely calls to mind the Ableton "Push" hardware controller for the company's *Live* software. As with so many musical interfaces, ingrained into the design of this device is an entire history of instrumental and gestural affordances borrowed from the drum machines, samplers, keyboards, and mixers that have emerged in the past century, reflecting what music software and digital media expert Thor Magnusson calls "ergomimesis."[2] The physical gestures and tangible metaphors inscribed into the hardware are mediated by the fluidity of the software, with its constant updates and alternative visual aesthetic that encourages shifting understandings of musicality and instrumentality. The device manifests a perennial struggle in the design and use of musical interfaces. As the affordances of software push against the embodied memory of hardware, ontologies and epistemologies of musicality, instrumentality, and sound itself are put into question. While music software is often pushed onto its users under the guise of accessibility and "user-friendly" design, and musicians often embrace the button pushes, mouse clicks, and knob twists that define digital media production, how might these interfaces push back against both software and user, apparatus and individual? As discussed throughout this book, the standards and conventions of software design are never simply technical but also reflective of shifts in social and cultural values, ideologies, and ways of thinking and being. *Push* describes how twenty-first-century interface design shapes musical practices but also how the seemingly separate processes of technical design and cultural use push against each other in the creation of, and in response to, broader social values.

In short, I'm interested in how software acts as a metaphorical interface to the formation of society and culture. By studying developments in the design and use of *sound* software, in particular, I detail three shifts that have occurred in music and interactive media production during the early twenty-first century. First, I trace a shift in conceptions of sound from *text* to *experience*. Through much of the twentieth century, sound recording technologies were designed for the purpose of audio inscription. From the phonograph to magnetic tape and the vinyl record, sound was conceived to be an archival text of sorts, faithfully preserved as it was etched into the media.[3] Thinking of sound in this way has led many practitioners down a seemingly never-ending quest for fidelity—constantly seeking out *media* that most perfectly reproduce sound in its impossibly *unmediated* state. From early Victrola advertisements that claimed "the human voice *is* human" on the new phonograph,[4] to the "crystal clear fidelity" of Memorex cassette tapes,[5] the history of sound media in the twentieth century was primarily concerned with capturing sound as an *object*.

By contrast, the emergence of digital tools for sound reproduction in the new millennium has encouraged conceptions of sound as an *experience* rather than an object. The marketing rhetoric of sound "immersion" has altered previous object-oriented fidelity discourses, as in the "4K" television ads that promote "absolute audio clarity" so perfect that you do not simply hear it but immerse yourself in it.[6] Conceptions of sound in the digital age have also become more abstract, dismissing entirely claims about sonic fidelity. Ambient musical soundscapes and the background sonic textures of video games, for example, define many of the media experiences discussed throughout this book. In contrast to the "foley" model of sound design, which strives for the faithful reproduction of sounds found in the "real" world, the most popular "indie" video games and mobile music production apps provide unimposing backdrops that allow users to explore virtual environments removed from the values of sound fidelity.[7]

In addition to the shift in conceptions of sound from text to experience, I outline a shift in the structure of media design from linear *narrative* to nonlinear, process-oriented *exploration*. This second transformation has to do with how users of a given media platform experience the medium itself, rather than the content or message being conveyed by the platform. In this context, "the medium is the message," following media theorist Marshall McLuhan.[8] For example, the advent of internet web pages and the "World Wide Web" in the mid-1990s introduced nonlinear mechanisms for reading text, such as browsers, hyperlinks, and markup languages. In contrast to the linear, goal-oriented structure of paper books, these "hypertextual" forms of web-based media were meant to be explored in a fragmentary manner, the readers allowing themselves to be distracted by the linked and networked nature of online content.[9]

The shift from narrative to exploration is equally apparent in the changing trends in music software design. As software companies compete by offering the biggest and best features available to the user, musicians develop new methods for navigating these affordances. While the vast amount of add-on "plug-ins," constant software updates, and capacities for connecting software with external applications can be overwhelming to some musicians, others embrace the maximalist nature of music production software as a platform for creative experimentation. In this context, musicians *explore* the interface rather than *use* it for specific purposes or intentions.

Finally, this project details a shift in the fundamental understanding of media from *object* to *environment*.[10] Similar to representational models of sound media, software design historically has privileged a fidelity between "real" and "virtual" objects.[11] The history of personal computing has encouraged a conception of the "interface" as the visual and physical surface of a device, simply meant as a direct control mechanism for the user. While the graphical user interfaces (GUIs) of software are often designed to give the impression of "direct manipulation" between human and computer, it's important to recognize that the digital buttons, knobs, and icons with which we interact on a regular basis are just virtual *metaphors* for physical objects.[12]

This "skeuomorphic" design philosophy, in which interface elements are metaphors for "real" objects, has become less prevalent with the advent of mobile media such as smartphones.[13] Instead, abstraction has become a guiding principle in GUI design, thus focusing the computer user's attention on the environment of the operating system as a whole. In direct opposition to the material, object-oriented metaphors that littered Mac and PC "desktops" throughout the history of personal computing in the twentieth century, the launch of operating systems such as Windows 8 and iOS 7 in the early 2000s ushered in an era of "flat" design.[14] Here abstract shapes and colors replaced desktop metaphors such as "trash" icons and "folders" as overarching design trends to guide the user experience (UX). Consequently, as screens got smaller and interfaces more abstract, the ways in which users interacted with them became more immediate, tangible, and visceral. Mechanical gestures such as "clicking" a mouse and "typing" on a keyboard were gradually replaced by touch-based gestures in which the user directly "tapped," "swiped," and "pinched" the interface itself. The embodied techniques afforded by mobile media thus highlight how "digital" software, despite its supposed ephemerality and immateriality, still relies on the broader hardware environment surrounding its usage.

As a result of the increasingly material design of software, in which virtual interface elements express the visceral affordances of touchscreen devices, conceptions of both sound and interface have become more fluid, moving away from representational, object-oriented models that seek fidelity.[15] Instead, sound and interface are increasingly understood as comprising an entire media environment that includes hardware and software technologies, as well as the cultural context within which these technologies circulate. Here the "Push" metaphor is useful as a model for understanding how software cultures simultaneously reify and push against Silicon Valley's neoliberal ideal of "design thinking," both embracing and rethinking concepts of "innovation," "play," "emergence," and "iteration," among others.[16] In the case studies presented throughout this book, I use these terms to reflect the dialogic relationship between software and culture in the first two decades of the twenty-first century and to offer a practical model for the hermeneutic and semiotic analysis of software design as a cultural process rather than a fixed "text."

Indeed, the social and cultural timeline of the 2010s is itself defined by dialogue, dichotomy, and tension. While the first half of the 2010s was marked by a seemingly boundless optimism toward new media (reflected in the Low End Theory and indie gaming scenes described earlier), the second half of the decade witnessed a critical "wokeness" in the entertainment industries, marked by the emergence of multiple social movements dedicated to increasing public awareness of issues such as racism and sexism in the music, film, and game industries. In 2013, the Black Lives Matter movement began as a hashtag on social media dedicated to protesting racially motivated violence against black people after George Zimmerman was acquitted in the shooting and killing of African American teenager Trayvon Martin in 2012. The rest of the decade was defined by political unrest and racial tension as a result of often unchecked police brutality, from the large-scale protests in Ferguson, Missouri, following the murders of Michael Brown and Eric Garner in 2014 to the nationwide, months-long civil unrest following the murders of Breonna Taylor and George Floyd in 2020.

Following high-profile sexism in the film, game, and music industries, "popular feminism" became a major social movement in the second half of the 2010s. In 2014, Eron Gjoni—ex-boyfriend of game developer Zoë Quinn—wrote a negative blog post about Quinn, which resulted in internet users falsely accusing Quinn of having an unethical relationship with video game journalist Nathan Grayson. The post sparked an online harassment campaign from so-called Gamergaters that included doxing (releasing personal information), rape threats, and death threats against Quinn, fellow game developer Brianna Wu, and media critic Anita Sarkeesian. In what became known as the Gamergate controversy, the stakes of the 2010s culture wars were made clear as the game industry was forced to acknowledge its severe lack of representation from women and people of color. The Me Too movement caused similar shake-ups in the film and music industries. First coined by activist Tarana Burke in 2006, Me Too spread on social media in 2017 in response to widespread sexual-abuse allegations against film producer Harvey Weinstein. In an effort to shed light on the magnitude of sexual harassment and abuse toward women in America, the movement encouraged survivors to post "Me Too" as a social media status. The movement gained widespread media coverage after many celebrities and Hollywood actors took part, leading to high-profile firings, controversies, and organizational cleanups across the entertainment industries.

A large part of this book was written during these movements. The tension between technology as an empowering tool for creative liberation, on the one hand, and, on the other, as a dangerous tool for large-scale surveillance, bullying, and all sorts of "isms" formed a core component of the cultural politics that I witnessed in Los Angeles across the digital cultures of music, games, and interactive media. While I don't claim to provide complete answers as to how music software companies have addressed, or plan to address, issues of diversity, equity, and inclusion, each chapter speculates, at least briefly, on how software design itself might be considered a by-product of the social and cultural politics of music production. Considering that most of the musical examples I address stem from traditions rooted in the African diaspora, Paul Gilroy's idea of the Black Atlantic as a symbiotic "counter-culture of modernity, forged through the diametric but symbiotic opposition between Western Europe and its colonial 'Others'"[17] provides a fundamental premise of the "Push" metaphor and a useful lens through which we can recognize the simultaneous empowerment and damage that traditional approaches to HCI and design thinking have done in terms of race, gender, class, and ability. Within the broader context of personal computing, for example, major issues in software design and distribution have included severe wealth gaps and gendered access limitations to computers at home, in schools, and in the workplace. As discussed in chapters 3 and 5, the history of computer science as a professional and scholarly discipline has been more welcoming and inclusive to young white men from privileged backgrounds than other students, regardless of skill level and professional aspiration. Further, interaction design has focused historically on a limited set of physical peripherals, interfaces, and interaction guidelines that favor largely able-bodied users.

Similar issues have plagued music software development and virtual instrument design. As discussed in chapter 2, digital audio workstations (DAWs) have for a long time favored skeuomorphic design and encouraged the use of Eurocentric musical standards such as the twelve-tone scale, C major key, 4/4 time signature, 120 tempo, and compositional affordances (the piano keyboard layout as the main compositional and performance interface, a linear timeline as the arrangement interface). In a *Pitchfork* article,

"Decolonizing Electronic Music Starts with Its Software," journalist Tom Faber writes about how "music production tools have the unconscious biases of their creators baked into their architecture."[18] In one of his featured interviews, electronic musician Khyam Allami claims that popular DAWs such as *Live*, *FL Studio*, and *Logic Pro X* were built to facilitate music making according to the Western standards and principles developed from European classical music, arguing that these programs force music into "an unnatural rigidity." For Allami, the music made by non-Western creators especially "stops being in tune with itself," and "a lot of the cultural will be gone" as a result of what he calls the current "colonial, supremacist paradigm. The music is colonized in some way."

This reflects a broader push to decolonize institutions that have fostered white supremacist and colonialist histories of music making. In her 2018 article "The Musicological Elite," musicologist Tamara Levitz exposes the white supremacist foundations of how the study of Western music history became institutionalized, as well as the racist backlash against diversity, equity, and inclusion initiatives presented by organizations such as the American Musicological Society and the Society for American Music in the late 2010s.[19] In 2020, music theorist Philip Ewell published the groundbreaking article "Music Theory and the White Racial Frame," in an attempt to decolonize the "white racial frame" that has structured the history and practice of music theory and therefore a large majority of compositional techniques and forms of musical analysis that have defined Western classical and popular music.[20] Ewell's piece was especially controversial, leading a group of music theorists to publish a "special issue" of their academic journal that focused on disputing his ideas using the same racist ideas he was trying to expose. In response, the journal faced calls to cease publishing.[21] While the professional disciplines of musicology and music theory may seem like obscure bastions of elite, ivory-tower musicians, historians, and theorists, these debates reflected the broad impact of the Black Lives Matter movement, as well as the politically charged nature of free speech in the late 2010s. Many of the chapters in this book describe, at least in some capacity, how music software development has responded to these attempts at decolonizing music making in the Western world.

Similarly egregious in the context of music software development has been the lack of focus on accessibility features, especially for users with visual impairments. One notable exception has been Native Instruments, which introduced a robust "accessibility mode" to its Komplete Kontrol keyboards in 2016, but the extent to which its efforts have been replicated in the music products industry was unclear by the decade's end.

Finally, we might consider how major music software and instrument manufacturers have based their designs on the creative labor of black musicians, "emulating" musical devices and techniques to create profitable plug-ins and virtual instruments. This is a core example of the "immaterial labor" in which digital creatives often find themselves enmeshed, contributing physical, intellectual, and creative work to software companies under the guise of "collaboration" and with the promise of promoting their artistic careers. Ethnomusicologist Michael Veal claims this to be an inherent feature of black aesthetics in the African diaspora, citing dub reggae's remix culture as an example of Gilroy's "counter-culture of modernity."[22] Following Veal, *Push* examines hip-hop and EDM as traditions born out of both a resistance to American capitalism and to consumerism by black musicians and a strong embrace of and reliance on the tools of global capitalism in order to carry out their cultural politics.

Outside of music, the rise of the "military-entertainment complex" is an example of the problematic convergence between HCI, new media platforms, and the entertainment industries that has resulted from the increasing ubiquity of software. The 2010s witnessed developments such as video games for military training, the use of video game controllers in piloting military drones, and the integration of global surveillance systems with everyday technologies such as the smartphone. By balancing critical and practical approaches to software design, I engage simultaneously with the concrete creative practices of musicians and software designers, as well as the ideological and ethical questions surrounding HCI. Conceptualizing the dynamics of HCI as a relational process of cultural and technical negotiation becomes especially important as software continues to ingrain itself into the transnational machine of neoliberal capitalism.

Push expands on insights generated by scholars and practitioners in two subfields of music and media studies: sound studies and software studies. With the emergence of the interdisciplinary field of sound studies, scholars have pushed discussions of sonic engagement outside of strictly musical practices, asking questions related to the history and nature of sonic experience. Media theorist Jonathan Sterne, for example, provides a synchronous account of developments in media such as telephony, phonography, and the radio, as well as the modern practices of medicine, physics, and industrial capitalism, thus detailing the technological preconditions for the emergence of a more integrated sound *culture*.[23] Although Sterne is partly concerned with sound's *object-oriented* ability to be recorded, inscribed, and archived in various ways, he conceives of listening to and producing sound primarily as experiential *processes* that help in shaping ongoing relationships between humans and technology.

Indeed, sound has come to play a major role in defining relational processes between society, culture, and the natural world. In effect, sound scholars have conceptualized sound itself as an interface of sorts. Stefan Helmreich uses the notion of "transduction" to describe how sound metaphorically transmutes and converts as it moves across media, similar to how sound vibrations are transduced into energy by the inner ear.[24] Transduction describes more than just the physical conversion of energy forms. Rather, the concept characterizes the entire relational network surrounding the sonic interaction, from the technical design of the system (sound, image, and mechanics of the media interface) to the aesthetic and ethical understandings of sound at stake in the listener's experience.[25] Alexandra Supper and others have examined the social and cultural consequences of sonification technologies that render scientific data into sound, as well as the ways in which devices such as the stethoscope "audify" bodily phenomena.[26] Similarly, professional sound designers working in the field of sonic interaction design have developed strategies for using sound as a cognitive and embodied cue in the experience of non-"musical" media.[27] In each case, sound is a dynamic process of relationality *between* social, cultural, and technological agents, rather than a fixed consumer product that is controlled by a "user."

If the goal of sound studies is to expose the sociotechnical and historical infrastructures that shape the production and reception of sound, media theorists working under the rubric of "software studies" are interested in pinpointing the technical structures that shape the design and use of software. In *10 PRINT CHR$(205.5+RND(1)); : GOTO 10*, Nick Montfort and coauthors use a single line of code from the Commodore 64 computer as an analytical lens in order to consider the cultural impact of software.[28] Lev Manovich's *Software Takes Command* examines media production software such as the Adobe *Creative Cloud* to highlight how software techniques become established creative practices across a range of cultural communities.[29] Following in the footsteps of classic media formalists such as Friedrich Kittler and Marshall McLuhan, software studies scholars engage in hermeneutic and semiotic analyses of software programs to answer more expansive questions related to the social, political, and economic effects of software, as well as how changes in software design affect cultural practice.[30]

Foundational texts in software studies deal broadly with the relationship between software programs and cultural practice—that is, the interface between technical *codes* and their realizations as social *forms*.[31] Scholarship in "interface criticism" extends the focus from the inner workings of software to the more user-oriented elements of HCI, the interface itself. Summarizing what he terms "the interface effect," Alexander Galloway states, "the computer is not an object, or a creator of objects, it is a process or active threshold mediating between two states."[32] This relational ontology has become the dominant theme in "interface studies." Media theorists Branden Hookway and André Nusselder frame HCI through the positionality of "in-between-ness," in which the foundational experience of technological mediation comes from the active process of mediating between the two states of reality and virtuality.[33] While interface studies serve as a crucial body of research in highlighting the ontological and phenomenological experience of mediation, it has little to say about the formal and aesthetic aspects of technological experience in practice, as well as the multimodal experiences encouraged by "interactive" music and sound.

In a similarly critical gesture, design research professionals have increasingly shifted their language from the object-oriented nature of *interface* to the process-oriented nature of *interaction*. Designer Brenda Laurel claims that in focusing on what the *computer* is doing, "interface designers are engaged in the wrong activity." Instead, she suggests that we throw out the term "interface" altogether, since the main focus of media design should be "what the *person* is doing with the computer—the action."[34] Laurel's work, along with that of pioneering design guru Donald Norman, has had a direct influence on the emergence of UX and interaction design as disciplines that understand media interfaces as process-oriented "experiences" rather than material products.[35] Scholar and practitioner Janet Murray expands on this work by acknowledging the significance of the social and cultural context in any technological interaction. For Murray, "interactors" (rather than "users") are engaged "with one another through the mediation of the machine, and with the larger social and cultural systems of which the automated tasks may only be a part," thus reminding designers that the mere "usefulness" of an interface is less important than its context.[36] Together, these developments in scholarship and professional practice encourage changing conceptions of the term "interface" itself.

In contrast to *object*-oriented definitions, I suggest thinking about interface as a *process* that encourages playful forms of creativity, the convergence of tools and techniques across media, and a relational ethic of technological interaction. I call this method *interface aesthetics* because it allows us to get beyond current understandings of technological interaction as being rooted in either the experience of user control over technological materials or the technical underpinnings of the material technology itself. While prior uses of the term simply referred to the visual look and feel of interface design, this book combines the process-oriented definition of "interface" with a more holistic, contextualized definition of "aesthetics," referring to how software users negotiate new forms of technological interaction as part of their social and cultural identities.[37] A fundamental premise of this book is that by analyzing simultaneously the design and use of software, it's possible to reveal the social, cultural, and ideological structures embedded within the aesthetic and technical design of media objects.

Coined by German philosopher Alexander Baumgarten in 1750, the term "aesthetics" aligned with the original Greek word for "sensory perception," dealing with not only art *objects* but the *process* of human perception in its entirety.[38] While philosopher Immanuel Kant's notion of the term as a "disinterested" withdrawal from social, political, and bodily dictates of value judgment remains the most influential conception of aesthetics, interfaces complicate this objective understanding of the term.[39] On one hand, the use of human-computer interfaces is an intimately visceral and sensual practice that foregrounds the inner structures and mechanics of the technological system. Arguably, the main reason an interface such as the computer keyboard has been successful lies in the intuitive ways in which it externalizes the text-based nature of digital code, making tangible the ephemeral nature of software. On the other hand, many interfaces are designed to hide these very structures through flashy screen layouts and novel interaction patterns. Mobile media interfaces, such as the iPhone touchscreen, are designed to foreground the physical gestures of the device's user rather than the "rules" of the software code. As such, interfaces are far from "disinterested" agents in the shaping of aesthetic judgment and sensory perception in the digital age.

Since the interface plays such a pivotal role in the shaping of contemporary aesthetic values, the pervasive nature of interfaces in the early twenty-first century calls for a similarly integrated definition of aesthetics. Throughout this book, I follow Terry Eagleton's conception of the term, recognizing that to judge aesthetically is to compare values and that those values emanate from the totality of the judge and their context.[40] In this way, aesthetic judgment is at once a political and ethical act capable of radically transforming existing conceptions of art, culture, and society. We might also recognize that social, cultural, and political practices are always transduced into technical design, and vice versa.[41]

My research asks this question: as sound continues to become enmeshed in software, how do sound artists adapt existing embodied practices to constantly changing interfaces? Moreover, how has the emergence of alternative interfaces ushered in new conceptions of sound in the digital age? Building on musicologist Nina Sun Eidsheim's work on sound as an intermaterial and relational process, I understand sound *software* as an interface for "vibrational" transmissions that connect computer users to their social, material, and cultural contexts.[42] The idea of sound as a relational *process*—that is, as interface—introduces the more drastic notion that music is no longer "Music" as it has been defined traditionally but is a node within a broader media ecology. Indeed, as sonic ontologies increasingly become digitized, shifting from objects to processes, listening itself becomes rewired. As the examples from Low End Theory and indie gaming make clear, these ontological shifts occur in parallel with new cultural formations. In this context, consumers of digital music and media are not just passive listeners, but they approach artistic consumption from the perspective of makers themselves. I refer to this mode of HCI as *procedural listening*.

In order to examine the interrelated and mutually dependent relationships between software design and the creative practices surrounding digital audio production across media, *Push* engages multiple methodologies, including print and online discourse analysis, analyses of software and hardware devices, participant observation, and insights from practice-based research, through my work as a hip-hop producer, EDM DJ, music technology instructor, and digital media artist working with sound. These sets of inquiries have been defined by the contacts and collisions between bodies and material technologies, connections and disconnections between technological devices, and the fluid web through which tools, techniques, and practices spread when sound is encoded in software.

I chose these methods based on my fundamental position that new insights can be gained by thinking through the coterminous development between forms of analysis and forms of creative practice. Ultimately, the convergence of technical analyses, ethnographic insights, and practical experience constitutes the major contribution of this project. In contrast to ethnomusicological or anthropological monographs which use ethnographic methods to describe the practices of individual people or cultural communities, this book employs participant observation methods for the purpose of understanding broad, collective knowledge surrounding digital media production in the first two decades of the twenty-first century. As a result, most of the "data" presented in the text come from formal, hermeneutic, and aesthetic analyses of software rather than direct quotes or ethnographic field notes. The insights generated from these analyses are supported by the collective insights gathered through more than twelve years of creating, performing, and working with the cultural communities under discussion.[43] The pervasive emphasis I place on *design* as an intermediary between production and consumption is meant to highlight further the process-oriented nature of the term "interface aesthetics"—a general framework for thinking through relationships between theory and practice, concept and technique, aesthetics and poietics. Of course, in the analysis of cultural objects as dynamic and complex as software, there are benefits and limitations to each individual research method.

Hermeneutic analyses of software applications provide the bulk of my research data. The technical practice of designing possibilities for human interaction with technological systems has been broadly defined, with subfields including user interface (UI) design, HCI, and UX, among others.[44] I develop a methodology for analyzing interface design that combines multiple practices within these fields. First, I examine the visual design, or "information architecture," of software and hardware interfaces, focusing on how the organization of buttons, knobs, sliders, and other control mechanisms creates a system of affordances and constraints related to the usability of a given device. Then I detail how the visual form of the interface—combined with the "back end" logic of the software code—suggests certain models for sonic and compositional form. Finally, I consider how the design of specific interface models encourages novel forms of performance. Together, these analyses broadly define the information architecture of contemporary interfaces. For designers, this approach may enhance the critical power of the work with which they're engaged. For scholars, this approach offers a framework for understanding the relationship between the creation of a work and its broader cultural context—a type of "sketch studies" for the digital age.[45]

Discourse analysis is equally useful in ascertaining the connections between *how* producers technically engage digital audio tools (poietics) and *why* particular practices become standardized over others (aesthetics).

I examine the marketing materials for various music software; debates taking place in web forums, Facebook groups, Twitter lists, and other on-line communities of audio producers; and pedagogical materials, including YouTube video tutorials and trade publications. With the rise of what eth-nomusicologist Kiri Miller calls "amateur-to-amateur" learning, marked by an increasing proliferation of both tools and techniques through vir-tual networks, discourse analyses allow me to detail emerging trends, rhet-oric, and aesthetics of audio production across communities of practice.[46] Furthermore, the combined approaches to discourse analysis from Michel Foucault and Friedrich Kittler offer useful paradigms for exposing the ideo-logical conditions that give rise to various media practices.[47]

An ethnographic approach further grounds my technical and discursive methods, and Southern California—what many hail as "the next Silicon Valley"—has been a particularly fertile ground for this project.[48] Participant observation within various electronic music and digital media scenes—fo-cused in Los Angeles and extended virtually through my online network of DJs and producers—influenced the bulk of my research insights, as I attended and participated in countless music performances, game festivals, technology conferences, and media workshops while living in Los Angeles from 2011 to 2018. These close observations offer unique insights into the public performances and communal aesthetics of various subcultures. While I interviewed a range of software developers, electronic musicians, sound designers, game designers, and other digital creatives, I chose not to ethnographize these data formally because most of the insights gained in this way focused on the informants' personal or company-based marketing efforts and commercial goals. There are plenty of moments throughout *Push* in which the voices of software's users could be beneficial, especially in un-derstanding power dynamics of race, gender, class, and ability discussed in each chapter. Ultimately, I chose instead to let the work of these individuals and companies speak for itself by pointing to the outcropping of public and documentary media in which their voices are heard in a much more direct and immediate way than an academic book could ever get across. Part of the radical, if speculative, nature of *Push* lies in the belief, following Henry Ford, that "all objects tell a story if you know how to read them."[49] If we can learn how to "read" software programs critically, we can get one step closer to bridging the perceived gap between how technology is designed and how it's used. In the end, this combination of macro and micro approaches builds on anthropologist Sherry Ortner's practice of "interface ethnography," in that it deals not only with events and practices in which a subculture *interfaces* with a given public but also with how the technological design *interfaces* with both technical practices of audio production and aesthetic aspects of cultural formation in the digital age.[50]

Finally, this study is unavoidably shaped by my perspectives as a music producer and music technology instructor navigating constantly shifting trends in sound, software, and digital audio. Since 2005, I've produced music and sound art as a hip-hop beatmaker; a "chipmusic" artist who circuit bends, recombines, and performs with vintage computer technologies; a sound designer for games, VR (virtual reality), and interactive media; and a DJ. At the same time, I've taught these tools to students in higher education across music, media studies, music industry, and computer science departments as an Avid Certified Instructor and an Apple Certified Pro in *Logic Pro X*, working with students to navigate constantly changing interface designs and routinely troubleshooting software to optimize creative workflow. The idea of thinking *through* practice—or what media theorist Peter Lunenfeld describes as "the maker's discourse"—has gained traction across disciplines, from new media theory and the digital humanities to musicology.[51] In her foundational work, *Boccherini's Body*, cellist and music historian Elisabeth Le Guin describes how the kinesthetic, proprioceptive, and tactile aspects of musical performance encouraged a revelation in her methods of musicological analysis: "even as the centrality of a visual listening was becoming evident to me, I was increasingly convinced that certain qualities in Boccherini's music were best explained, or even solely explicable, through the invisible embodied experiences of playing it."[52] In coining the term "cello-and-bow thinking" to describe a mode of analysis that attempts to draw affective meaning from nuances in the physical gestures and muscular distribution of musical performance, Le Guin provides a useful phenomenological method for scholars attempting to combine theoretical and technical aspects of musical practice.[53]

Rather than providing a cultural and technical overview of the historical development of software, *Push* instead presents a specific transitional moment in the first decades of the twenty-first century in which software became the primary platform through which artists engaged with sound and media. The book is organized into three parts. Part I, "Sonic Architectures," discusses how the design of music production software reflects broader cultural aesthetics. Part II, "When Hardware Becomes Software," examines how the affordances of music software are materialized in both live performances with technology and everyday embodied interactions with hardware such as the iPhone. Part III, "Software as Gradual Process," discusses how the emergent aesthetics of artificial intelligence in video games and music mixing software suggest more dynamic relationships between the technical aspects of software and the aesthetics of UX. Ultimately, the book traces the conceptual shift from a tool-based understanding of HCI to a process-oriented and relational awareness of digital interfaces as active agents in the shaping of ethical forms of interaction with technology.

Part I of the book, "Sonic Architectures," deals with how the GUIs of music software programs shape conceptions of foundational musical concepts such as composition and instrumentality. Like that of most software in the history of personal computing, the design of music production software has been guided by a dichotomy between "usability" and "hackability." On one hand, there are popular commercial software programs that simulate existing tools and techniques in order to make the application more user-friendly to a broader demographic of consumers. On the other hand, there are open-source, and often free, software programs whose source code is public and open to custom modifications from a community of "do-it-yourself" programmers. These disparate approaches to software development entail broader socioeconomic values, political ideologies, and cultural aesthetics. By analyzing the contrasting approaches to GUI design in software such as *FL Studio*, *Pro Tools*, *Live*, and *Max*, chapters 1–3 reveal the values, aesthetics, and ideologies fixed within music production software.

Chapters 1 and 2 focus on what are arguably the most popular music production software programs introduced to date: Image-Line's *FL Studio*, Avid's *Pro Tools*, and Ableton's *Live*. In the mid-2000s, *FL Studio* and *Live* became the dominant software programs for electronic music producers working across genres. Designed for "real-time" musical performance onstage and in the studio, *Live* introduced unique affordances for engaging digital sound, such as a nonlinear "Session View" which allows for the modular juxtaposition of musical ideas, as well as increased interoperability between *Live* and other media. Through design analyses of the software interfaces of *FL Studio* and *Live*, as well as discourse analysis among producers and designers, chapters 1 and 2 detail how the design of music software is moving away from the representation of preexisting instruments toward more abstract interfaces for engaging digital sound. I contextualize the aesthetic desire among producers to integrate a vast array of media and technologies into their creative workflow within the broader concept surrounding consumer technology known as digital maximalism—a philosophy that says that the more connected you are, the better. In doing so, I answer fundamental questions regarding the multimodal nature of musical composition with the emergence of software.

Chapter 3 aligns the practices of musical composition in *Max* and *Pure Data* with the technical skills and design aesthetics of computer programmers to outline a theoretical model for computational thinking in electronic music. While *Live* has appealed to popular music and EDM producers because of the influx and accessibility of creative options presented by the interface, other music software embodies minimalist aesthetics, fostering creativity through limitations in design. *Max*, for example, is rarely described as software at all but rather a musical "environment." As the visual and algorithmic nature of the program suggests, *Max* requires composers to embrace an alternative digital literacy when working with the software, one more akin to computer coding than to traditional music composition. Outlining perspectives on the practice of coding from software studies, computer science, and the individual aesthetics of software developers—including those of original *Max* developer Miller Puckette—I present insights into the nature of composition and instrumentality in the digital age. Most noticeably, computational thinking encourages artists to think increasingly through the lenses of designers, crafting entire systems rather than individual "works" and developing proficiencies in media formats and techniques that continue to converge in the screen space of software.

Part II of the book, "When Software Becomes Hardware," examines how hardware controllers have allowed electronic musicians to physically embody the gestural affordances of software, thus establishing a feedback loop between the body of the performer and the computational system. In recent years, touchscreen technologies have become the dominant modes of engagement with digital media, particularly in the realm of music production. The rise of mobile apps for creating music, kinesthetic control schemes in music-based video games, and an increasing abundance of hardware peripherals for controlling sound in DAWs have coincided with the rise of "accessibility" and "tangibility" as rhetorical metaphors of control in interactive media. While digital audio production tools are often marketed for their "democratizing" capabilities, it is exactly through this rhetoric of accessibility that these "controllers" make sense in the era of digital convergence. "Touch" becomes a tangible metaphor for the desire for non-mediation and connectivity that mobile social media, video games, and digital audio production strive for but continuously fail to achieve.

Chapter 4 presents a case in point. Throughout the history of hip-hop and EDM, the process of mixing and manipulating vinyl records between two turntable decks has become standard practice for DJs, imbuing the performance with a sense of improvisational spontaneity and enhancing the perceived "live" presence of the DJ. However, the increasing presence of digital software in the DJ booth has thrown into question the nature of performance within EDM communities. To heighten the sense of physicality and direct manipulability when working with seemingly intangible software, producers and DJs have increasingly integrated button-based hardware "controllers" into their creative workflows. Combining design analyses of hardware devices such as the Monome "grid" controller and Ableton's "Push" with analyses of performance techniques from popular electronic musicians such as Daedelus, chapter 4 posits the shift from "turntablism" to "controllerism" as exemplary of trends toward the emergence of a "controller culture" more broadly.

Chapter 5 examines the Apple iPhone as a device whose user-friendly design and seemingly intuitive touchscreen mechanics equalize the skill levels needed for both everyday productivity tasks and creative music production. In contrast to controllerists who use hardware controllers to distinguish themselves from commonplace users of technology, iPhone users celebrate the disintegrating distinction between expert "producers" and nonexpert "users" facilitated by mobile media devices. The line between creative production and media consumption is further blurred with the integration of the *App Store*, an online shop where users can purchase both new apps and add-on content for existing apps on the iPhone. In analyzing how mobile media software and app design "democratizes" music production practices, it's possible to understand music more broadly as not simply involving *material* technologies (instruments, controllers) and traditional performance spaces (dance clubs, concert halls) but also as a process-oriented *experience* that aligns with consumption practices inherent in capitalism in the early twenty-first century.

Part III of the book, "Software as Gradual Process," examines the emergent and procedural aesthetics in video game audio and music mixing software, as captured by the following quote:

Things which grow shape themselves from within outwards—they are not assemblages of originally distinct parts; they partition themselves, elaborating their own structure from the whole to the parts, from the simple to the complex.[54]

I read this description of organic emergence, by philosopher Alan Watts, not while perusing treatises on Zen Buddhism or scanning self-help manuals for spiritual guidance but rather in the opening sentences of artist Matt Pearson's practical guide to using the *Processing* programming language for creating digital art.[55] The conflation of computer code and algorithmic processes with organic and holistic metaphors is commonplace in many contemporary digital art scenes and has a long history going at least as far back as the multimedia experiments of the 1960s avant-garde. How have the aesthetics of what has been broadly labeled "generative media" affected forms of digital audio production more generally? The final chapters of *Push* hint at the ethical implications of examining software as a process-oriented "experience" rather than a fixed commodity.

Chapter 6 examines the application of generative music to emergent media experiences, thus introducing new forms of HCI—what I define as *procedural interfaces*—through an examination of popular indie games such as *Proteus* (2013) and *Fract OSC* (2014). In a 1996 talk, "Generative Music," Brian Eno described the principle that formed the basis of his philosophy of ambient music: "the idea that it's possible to think of a system or a set of rules which once set in motion will create music for you."[56] With the rise of procedural generation in video game design and other forms of computer-generated media, this desire for self-generating "environments" would seemingly materialize in the multisensory space of video games. Extending a historical lineage of "generative aesthetics" throughout the twentieth century, I define the concept of emergence both technically and discursively, analyzing practices of sound design in the *Unity* game design software in conjunction with the theoretical and aesthetic motivations of game and sound designers themselves. Expanding on Eno's idea, procedural interfaces encourage dynamic, relational modes of technological engagement in which sonic interaction design guides the players through the virtual world, rather than a set of rules imposed by the designers themselves.

Chapter 7 thinks through debates in the late 2010s centering on the use of artificial intelligence techniques in mixing and mastering software. For decades, mastering has been defined by both its mystery (as the "dark art" of audio engineering) and its science (as the engineer learns to "master" the most obscure and fine-tuned details of acoustics and digital audio). For a profession so reliant on the perceived nuance of human listening capacities, what happens to the role of the mastering engineer in the age of big data, "intelligent" audio assistants, and the deep learning/listening algorithms of modern computing and social media networks? While the looming automation crisis forms a core premise of the chapter, I'm mostly concerned with how the diagnostic design and ocular-centric GUI of "assistive audio technology" software such as iZotope's *Ozone* encourage a biopolitical mode of listening that blurs the line between a musical panopticon and an empathetic form of HCI. In restructuring previous hierarchies of computational agency, these tools ask the big-picture questions about music's ontology in the late digital age: what are we listening to, what are we listening for, and who's doing the listening?

In the end, each chapter provides a case study in what I call procedural listening, a form of technological interaction in which the user focuses on the process-oriented mechanics of the technological system rather than the content created by the user. In facilitating a relational, two-way dynamic between process-based systems and human interactors, the environment provided by software encourages users to be more than just *users*: it inspires them to be *makers*. Yet, while many have rightly celebrated the participatory, "democratized" nature of digital media, the situation also calls for an increased attention to the ethics of HCI and an awareness of social responsibility on the part of both software designers and users. In this way, the fundamental goals of *Push* are both critical and practical. For scholars and researchers, the book aims to provide both a framework for analyzing software design as a cultural practice and a critical understanding of the increasing role software continues to play in the social, economic, and cultural lives of Western consumers. For designers and digital media artists, the book offers suggestions for moving beyond traditional creative techniques and practices, experimenting with new software interfaces as critical tools in themselves. By the time a reader finishes the book, a whole new set of software programs, digital interfaces, and media platforms will have been released, simultaneously extending and subverting traditional forms of creativity. Amid today's perennial upgrade culture, learning how to critically navigate emerging media and technologies becomes more relevant than ever.

PART I
SONIC ARCHITECTURES

1

Plug-in Cultures

On the rooftop pool deck of Drai's Las Vegas nightclub, club-goers pay the social media influencer known as "The Slut Whisperer" to pour thousand-dollar bottles of champagne over their bodies so they can get tagged in an Instagram post. Off the Strip, EDM DJs with multimillion-dollar club residencies such as Steve Angello of Swedish House Mafia blow off steam by shooting automatic machine guns at the local range, followed by test driving hypercars at two-hundred-plus miles per hour at the Las Vegas Motor Speedway. In the penthouse suites of the Strip's most luxurious hotels, *Forbes*-list DJs Tiësto and Steve Aoki discuss the rise of the DJ-entrepreneur and their plans for world domination.

These scenes come from a 2016 episode of Vice TV's *Noisey* music documentary series, and for many who have visited Sin City, they represent the apex of American greed, excess, and waste.[1] Throughout the episode, both the financial model of Vegas in the 2010s and EDM culture itself are positioned as products of the 2008 economic recession. Superclubs such as Hakkasan, Omnia, and XS tied their identities to the brands of top EDM DJs such as Diplo and Calvin Harris, and—in true Vegas fashion—people started paying a whole lot of money to appear wealthy in front of crowds of people they'd likely never see again.

While a maximalist aesthetic and neoliberal celebration of the EDM lifestyle are clearly evident in the countless scenes of conspicuous consumption throughout the documentary, it was a much quieter and more private scene that stood out to me. During a breakfast interview in the DJ's hotel room, *Noisey* host Zach Goldbaum asks Dutch DJ Afrojack about his creative process. Opening *FL Studio*, his DAW of choice, Afrojack starts making noisy synth sounds with his mouse and keyboard, explaining how with this software, he's made most of his biggest hits in a matter of minutes: "I open up a song that I made, and I use all the elements of that song to create another song. . . . You see how easy it is." Angello echoes this practice of viral music production elsewhere in the documentary, claiming, "Modern music, with the digitalization of it, has become this port of shooting up as much music as

Push. Mike D'Errico, Oxford University Press. © Oxford University Press 2022.
DOI: 10.1093/oso/9780190943301.003.0002

you can as fast as possible." ⓑ Digital media is often blamed for the downfalls of Western civilization, but really, what does a ninety-nine-dollar piece of software have to do with the conspicuous consumption and faux-lavish lifestyles being pursued in post-recession America? This chapter views the *FL Studio* music production software as a microcosm of the specific convergence of digital maximalist aesthetics, toxic masculinity, and neoliberal capitalism that defined EDM culture in the 2010s.

The rise of *FL Studio* parallels the viral spread and Wild West nature of internet culture in the late 1990s and early 2000s. Originally called *FruityLoops*, the software was developed in 1997 as a side project of a coder working for Image-Line, a Dutch-Belgian video-game company known for its adult-themed version of *Tetris*, called *Porntris*. The earliest iteration of the software consisted of only a four-channel step sequencer with a 127-note musical instrument digital interface (MIDI) bank to select a drum sound and a pattern selector that allowed the producer to choose from among nine different rhythmic patterns. Didier Dambrin, *FL*'s primary developer, calls the software "an app for cheaters," designed with a "visually appealing" display, intended to be fun, simple, and not much beyond that.[2] Echoing Afrojack's review from earlier, countless hip-hop and EDM producers from Soulja Boy to 9th Wonder and Porter Robinson attest to the rapid-fire pace with which they can produce music in *FL*. Image-Line founder Jean-Marie Cannie claims that the company "tricked people into thinking they have musical skills" through a simple-to-use interface that required little time, money, or effort to master.[3]

In its earliest iterations, the user-friendly nature of the software was reflected in a minimal interface that borrowed the core elements of analog drum machines and sequencers. The sixteen-step sequencing grid of machines such as Roland's TR-808 were (and still are) mirrored in *FL*'s gray and red step sequencer; patterns and tempo settings could be adjusted with simple up and down arrows; and presets were chosen from a matrix layout designed in a similar manner to a touch-tone telephone (figure 1.1). Importantly, the software completely ignored audio recording and editing capabilities, as well as any connection between MIDI note numbers and Western classical notation.

As the software matured, its status as a quick, user-friendly, hit-the-ground-running digital tool remained. The software contains a huge number of stock plug-ins out of the box, including sound generators, audio effects, and other plug-ins that reduce the need for third-party virtual studio technology (VST)

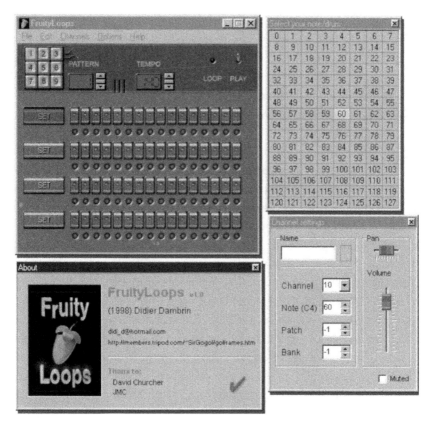

Figure 1.1 *FruityLoops* v1.0 (1998)

Source: https://www.image-line.com/history/.

or hardware instruments. Within music production circles, plug-ins such as the "Soundgoodizer" have generated heated debates about whether *FL*'s tools make music production so easy that using them might even be considered cheating.[4] Outside of audio composition and editing, the vast number of plug-ins also allow users to share projects for collaboration and bounce their audio tracks in a more streamlined manner.[5] Unlike other DAWs that include multiple compositional windows (Avid's *Pro Tools* "Mix" and "Edit" windows or Ableton's *Live* "Session" and "Arrangement" views), *FL* condenses micro sketch-based musical ideas with long-form arrangements into a single "Playlist" view. Third, the *FL* mixer panel is more open-ended in terms of the audio signal flow than the more isomorphic representations of channel strips in *Pro Tools* and *Logic Pro X*—audio inputs are assigned more freely without using the standard technical language for I/O (buses, e.g.), and sidechaining controls and other parameters are displayed in a more transparent way on the front panel of each track. Finally, *FL* includes randomization tools such as an automatic "Riff Machine" MIDI tool, complete with a "Throw Dice" button that can generate chord progressions, create chord progressions from existing musical ideas, and add performance variations to existing musical ideas (figure 1.2). ▶

From the start, though, it wasn't simply the user-friendly nature of the software that made it appealing to musicians and non-musicians alike. Like many startup software programs at the turn of the century, *FL* garnered its early reputation through illegal online piracy and peer-to-peer (P2P) file-sharing networks, or what became known as the "warez scene"—an almost exclusively male-dominated branch of hacker culture focused on the illegal distribution of software applications, or "warez."[6] In an online chat room from the mid-1990s, one user asks, "What is the link between the old days of BBSs, USENET, AOL Warez and today's file-sharing?" to which another user responds, "Young, adolescent boys, with a curiosity to learn about technology and sense of adventure. All these groups have been responsible for obtaining and cracking if necessary movies, software, and everything else. The thing you always have to keep in mind with all of these groups is that nobody really in the warez scene made money from the warez scene."[7] While

Figure 1.2 *FL Studio* v20 (2018) (screenshot taken by author).

FL Studio is now Image-Line's only product, Cannie argues that the company would have folded early on due to piracy if it hadn't invested time and energy in other software applications: "Thousands of people used it, but not a single soul was paying for it. We had [also developed] financial software, web development software . . . we had to do it to keep FruityLoops alive, because there was virtually no income."[8]

Typically, the warez scene is viewed as a radical subculture that embraced "anti-economic behavior" within a "modern gift economy."[9] However, the story of *FL* also reveals how warez were leveraged by software companies as sort of trial-run marketing strategies in an attempt to hook users into the app eventually. In the case of *FL*, internet piracy functioned as a form of neoliberal free labor—a partnership between software companies and their users built for the purpose of maximizing profit, while leveraging misguided libertarian ideals about the invisible hand of the internet. Like Vegas, internet culture and the DIY (do-it-yourself) music production market were built on the American dream that anyone can get rich off a single software program, mobile app, or chart-topping banger.

The history of the music products industry tells us that this has never been the case and that the "democratization" rhetoric about instruments being easy to use and accessible to all are mostly just marketing strategies used to encourage a maximalist consumption of technology. Music critic Ryan Diduck writes about how player pianos and other "easy-to-play" instruments in the early twentieth century didn't require the same time and effort to learn as more conventional instruments and therefore exemplified—like contemporary DAWs—"America's nascent consumerist mythology" which involved both "the maximization of leisure time" and "instant gratification."[10] Detailing how digital music production, specifically, involves more visual and less sound-specific practices, media theorist Ian Reyes claims that software like *FL Studio* has led to a "deskilling" of labor, as well as a "hunt-and-peck music production" style—a typing metaphor used to describe untrained computer users.[11] The hunt-and-peck metaphor is apt and reflects the constant searching for, and purchasing of, sounds required by the contemporary production process. Audio developer Mike Daliot argues that digital music producers "experience an overkill of gear use, which reminds me of my own lost years in chasing a perfect snare sample."[12]

However, despite the language of accessibility and empowerment surrounding music technology early on, the long history of the music products industry throughout the twentieth and early twenty-first century is marked by an increasingly homogenized consumer base, as well as constant attempts to make music technology more masculine. Musical instruments have been coded in gender-specific ways throughout the history of the music products industry, from the overtly phallic visual imagery of the electric guitar in the hands of male rockers to EDM DJs controlling and manipulating the technical apparatus of a full-frequency-spectrum sound system. As archivist Karen Linn notes, "A musical instrument is more than wood, wires, and glue . . . the essence of the object lies in the meanings the culture has assigned to it."[13] How might we read software in this way? Specifically, how might the design affordances of one of the most popular DAWs in the world mirror cultures of toxic masculinity in EDM and popular music more generally?

FL stakes its claim as a "democratizing" tool primarily in how it affords maximal accessibility to both the UI and the sonic material, with a minimal need for traditional audio engineering expertise. The software's multi-windowed "open" layout focuses on a multi-tier "toolbar"-style menu structure reminiscent of common productivity tools such as *Microsoft Word*. As such, the program emphasizes its maximalist, Swiss army knife–style affordances that encourage loose tinkering rather than a focused creative workflow (figure 1.3). ⊙ While this design layout has become a point of critique in software like *Word*, the idea is that this all-in-one tool can be picked up relatively easily by most users. In addition, the software uses a loose signal-flow taxonomy, both in the context of audio mixing and in the arrangement of musical ideas. In contrast to the stricter taxonomy and hierarchical relationship of "tracks," "channels," "buses," and so on, in "professional"-centered DAWs such as Avid's *Pro Tools* and Apple's *Logic Pro X*, *FL* is relatively agnostic to the relationships between sounds, instruments, and their positions within a musical mix. For example, individual "Tracks" in *FL* can contain "Patterns" of any different musical idea. A bass guitar pattern can be sequenced in the same track as a synthesizer pattern or drum loop, and each sound within each pattern can be routed to a "Target Mixer Track" without worrying about track groups, buses, or the traditional language of "send" and "return" tracks for effect routing. Finally, the software prides itself on its technological solutionist approach to stock audio effects, MIDI effects, and sound generators. Plug-ins

Figure 1.3 *FL Studio* v20 toolbar (screenshot taken by author).

such as the infamous "Soundgoodizer," "Scratcher," and "Effector" are single-knob interfaces meant to provide catch-all effects for any element in the mix, without the technical parameters of actual dynamics processors or time-based effects.[14] The "Soundgoodizer," for example, presents the user with nothing but a big knob and four lettered radio buttons that reflect different plug-in presets (figure 1.4). ▶

All that being said, how might we read these affordances as gendered? When you look at an electric guitar, both as a stand-alone object and the signifier of sexuality that it becomes when played by a male rocker, the phallic nature of the tool is obvious. With software, the connection isn't so clear. Historically, though, "software" has carried connotations of femininity, in contrast to its masculine counterpart, "hardware."[15] I argue that software like *FL* becomes masculine in how its "democratic" affordances are marketed and used in an effort to align music production with broader elements of control and mastery in neoliberal capitalism. Male professional and DIY "bedroom" producers alike often use the speed with which they craft compositions in *FL*

Figure 1.4 "Soundgoodizer" plug-in (screenshot taken by author).

as well as the quantity of their output as a testament to their masculinity vis-à-vis technical prowess.[16]

Given masculinity's attempt to define itself by control and mastery of reason, logic, and objectivity, it's no surprise that DAWs are often coded that way. Sociologist Judy Wajcman claims that "masculinity is allied to notions of technical competence, whereas women's apparent 'lack' of technical competence becomes part of a type of gender identity that is expected of women."[17] Sociologists Ann Game and Rosemary Pringle argue that it's necessary for men to maintain this "mystification of machines" in order to preserve not only male jobs but also the symbolic association of men's work with skills, as "the machine symbolizes masculinity and enables them to live out fantasies about power and domination."[18] Music software like *FL* has been particularly appealing to a mostly male consumer base, because the user-friendly nature of the interface affords this heightened sense of technical competence and control.

Here's the double-edged sword of democratization. On one hand, user-friendly design fosters the male fantasy of competence and control. On the other hand, democratization in design encourages broader demographics of users (women, children, those with less access to formal musical education, among others) to participate in music making. Especially during these moments when digital music production comes under threat of becoming a mass phenomenon, "gear lust" is often weaponized as a form of male technophilia to preserve the masculine coding of technology.[19] Starting in the early 1970s and increasing ever since, musicians' magazines have played a major part in encouraging the rapid consumption of new music technologies, often aligning new technologies with markers of both gender identity and expression, as well as new philosophies about consumption in neoliberal capitalism.[20] Marketing campaigns and user reviews for musical instruments often perpetuated the relationship between the male gaze and the desire for technical domination and control, conflating women with machines, the need for a human "feel," the call for increased control and expression with technology, and the presentation of seductive images of female sexuality as part of ad campaigns.[21] In a studio profile for *Pro Sound News*, Steve Harvey asks, "Who among us has not suffered from gear lust and the certainty that acquiring that one piece of equipment will, well, solve everything?"[22] In these examples, gear functions as an object of desire, a physical extension of the male user, and an intimate partner that provides services to the male musician.

Ironically, the consumerist mythology of speed, efficiency, and domesticity that has come to define music making with DAWs was perceived as a feminine quality of music technology in the early twentieth century. In fact, women amateur musicians played a huge part in making music with electronic instruments and providing technological innovations in electronic music within the home.[23] With the advent of the player piano, Diduck talks about how, "for women, music was domestic labor disguised as leisure. . . . For men, technologies were toys—infantilized, and feminized. . . . The toy metaphor reinforced the notion that electronic instruments were unproductive apparatuses of leisure and pleasure. It also created some commonly held notions: toys were cheap; toys required no skill; toys were for boys, not men."[24] Music blogger Orthentix highlights as one contemporary example of this a Facebook meme that displays a bunch of children's toys with the caption, "Dudes be like: 'I'm in the studio.' "[25]

In order to overcome the technology-as-toy metaphor and make music technology more masculine, the music products industry has had to transform constantly—through the crafting of institutional ideology, marketing campaigns, and highly gendered communities of practice—the image of electronic and digital devices from playthings to tools. Institutions such as Paris's Institut de Recherche et Coordination Acoustique/Musique (IRCAM) rationalized electronic and computer music by aligning creative workflows, organization of staff, and the technologies themselves with those used by large corporations.[26] Marketing campaigns for synthesizers and early digital instruments in the late 1970s and the 1980s often analogized music studio setups and creative workflows to the mundane tasks of white middle-class businessmen.[27] Cannie describes how in the early days of *FL*, Image-Line was "accused of making a DAW look like a toy when most competitors looked like gray accounting products."[28] He even hinted that the company eventually gave up the name "FruityLoops" partly because of its connotations of homosexuality.[29] Further, during the EDM boom and the heyday of DAWs in the 2010s, making music with software became a "guys thing," as countless internet memes and producer features in mainstream periodicals positioned the male producer as a mad scientist of sorts, hiding out in his man cave away from the feminized distractions of domestic life. Gendered divisions were often reflected in the subjects and imagery of the memes, which often used images of white women exhibiting carefree behavior as examples of "nonproducers" in contrast to serious white men as prototypical "producers" (figure 1.5).[30]

MUST HEAR Reference Songs
for Producers - Antidote Audio

Figure 1.5 Internet meme with gendered distinction between "producer" and "non-producer" (https://me.me/i/listening-to-music-before-being-a-producer-listening-to-music-bdf9948585c1455ab8aedd97c3181c54).

These implicit and explicit attempts at masculinizing music production software seem to work. Music education researchers have consistently found that both children and adults continue to associate specific instruments with corresponding genders, while others have argued that specific instrument designs favor the male body and that certain types of instrumental performances are gendered.[31] Victoria Armstrong's research on computer use in the music education classroom shows that male students often position themselves as more "expert" users than females and therefore have greater influence in shaping the classroom culture and curriculum.[32] Music composition itself has been associated with masculinity, as a discipline that "requires knowledge and control of technology and technique," and functions as a "metaphorical display of the mind," as music education expert Lucy Green writes.[33] As these examples demonstrate, "democratization" rhetoric within the music products industry has reified, rather than negated, inequity in gender dynamics and accessibility to music technology.

Perhaps it's best to read through the marketing rhetoric and understand that for the many tech-oriented industries focused on democratization, the goal isn't broad accessibility of use for their products but rather an indoctrination into a maximalist mode of conspicuous consumption and economic control in twenty-first-century capitalism. At the heart of this maximalist, male-centered techno-economic sensibility is software—the interface that provides users with a means for the everyday interaction and consumption of media. As musical instruments and tools increasingly converge in software platforms, this omnivorous approach to media consumption has broad implications for music composition and production.

Gear lust, masculinity, and the maximalist consumption of music technology aren't unique to the music products industry. They're both causes and symptoms of a broader maximalist aesthetic that emerged in the 2010s. In EDM, for example, music journalist Simon Reynolds describes the "rococo-florid riffs, eruptions of digitally-enhanced virtuosity, skyscraping solos," and other elements of digital maximalism.[34] His notion of both music consumption and production as driven by a "post-everything omnivorousness" reflects a shared economic sentiment with much digital maximalist discourse, highlighting the glutted mediascape and viral nature of neoliberal capitalism. For Reynolds, the blurry line between maximalist production and consumption is epitomized in the technical and practical process of digital audio production, in which there are "a hell of a lot of inputs . . . in terms of influences and sources, and a hell of a lot of outputs, in terms of density, scale, structural convolution, and sheer majesty."[35]

The "inputs" described by Reynolds aren't simply the wide-ranging artistic influences of the producer but also the hybrid, maximal nature of DAWs such as *FL*—software *environments* that foster the convergence of various musical tools into a coherent workflow and consequently shape the "texture-saturated overload" of the music. Music writer Matthew Ingram describes how DAWs encourage "interminable layering," the GUI presenting music as "a giant sandwich of vertically arranged elements stacked upon one another."[36]

The software's ability to tweak the parameters of any given sonic event results in what Ingram terms the "crisis of the control surface"—an overemphasis on the visual paradigm of software that "messes with a proper engagement with sound itself."[37] Echoing a common thread in maximalist discourse, Ingram perceives the inherently "wired-in" characteristics of software as deterministic properties that strictly prescribe and confine creativity, shifting the essence of musical practice from an organic "event" into an inscribed product. How might musical practice with DAWs such as *FL* make audible the sense of information overload inherent in the maximalist ethos in twenty-first-century digital culture?

The maximalist design of *FL* is most apparent in its multi-windowed "vectorial" workspace. Rather than flattening the program's tools into a hierarchical structure such as accordion menus or tabs, *FL* displays most elements of the interface—including the mixer, the step sequencer, the pattern arranger, the piano roll, and any plug-ins associated with the project—as separate windows layered on top of each other (figure 1.6). Across the windows, small icons are used to convey functions in addition to text breadcrumbs that denote a tab-style hierarchy. A variety of add-on windows provide extra functionality, including a MIDI touchpad pop-up and a script output window. To

Figure 1.6 *FL Studio* v20 maximal layout with touch pads, scripting, mixer, channel rack, piano roll, plug-ins, and sound library (screenshot taken by author).

help the user filter out all these options and create some semblance of information hierarchy, the "Project Picker" (figure 1.7) and "Plug-in Picker" (figure 1.8) windows create a graphical, network-style visualization of every musical pattern in the project, as well as all the possible plug-ins available to the user.

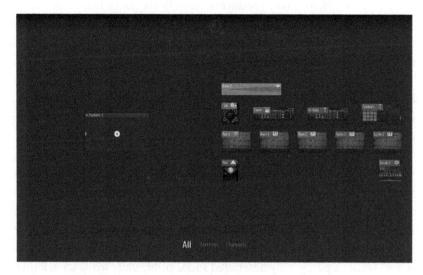

Figure 1.7 *FL Studio* "Project picker" visualization.

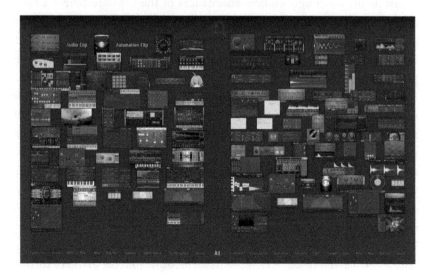

Figure 1.8 *FL Studio* "Plug-in Picker" visualization.

On one hand, this layout goes against conventional wisdom within UX and software design communities. The windowed desktop model mirrors early personal-computing operating systems which encouraged more directionless "browsing" rather than a guided workflow. On the other hand, the transparency afforded by the vectorial layout allows for easy access to library sounds and plug-ins, encouraging the rapid-paced workflow that has come to define *FL* for many producers. Jahlil Beats claims that he can knock a beat out in less than fifteen minutes, lauding the fact that "you can grab all the instruments at once and throw them right together on the grid and just click away."[38] Soulja Boy is even bolder in his praise for *FL*, claiming that "['Crank That'] probably took me like ten minutes to make, and everyone know I made like 10 million dollars off of the song. . . . People were like, 'Man, you used this demo version to make this song that went number one and made all this money.'"[39]

For so many producers, the accessibility of the software—its price, the number of sounds immediately available to the producer, seemingly endless functionality—is inherently tied to labor and profit. The sci-fi, flashy, maximalist design of the software analogizes the social and cultural arena of EDM itself: a fantastical space within the concrete reality of neoliberal capitalism, a space in which individuals feel empowered simultaneously to "be themselves" within a perceptibly unfettered social community while contributing invaluable labor to one of the wealthiest culture industries in the world.

Just as the seemingly endless affordances of the software reflect a rhetoric of democratization at the heart of technological design and use, *FL* also affects the cultural formations of DIY musicians and composers. Indeed, the apparent usability of the software has allowed it to become a ubiquitous presence in the everyday lives and personal spaces of electronic musicians. Dance music blog *XLR8R*'s "In the Studio" series takes the reader behind the scenes of popular musicians' creative workspaces, confirming the pervasive nature of software in the age of the "bedroom" studio (figure 1.9). However, while software developers and musicians alike praise the seemingly limitless affordances of programs such as *FL*, musicians often struggle to deal with the overwhelming number of creative options available to them in the instantly interconnected digital world. Scholars, journalists, and industry professionals have attempted to define this sensibility of information overload in digital culture, whether through the frames of technological "ubiquity,"[40] media "ecologies,"[41] or, even more broadly, epochal markers such as "the information age." None of these monikers captures the interconnections

Figure 1.9 Machinedrum in his bedroom studio, Berlin, 2012

Source: Shawn Reynaldo, "In the Studio: Machinedrum," *XLR8R*, August 27, 2012, accessed August 31, 2020, https://www.xlr8r.com/gear/2012/08/in-the-studio-machinedrum/.

between technical design and practical use in HCI. Instead, I suggest the term "maximalism" as an overarching concept that encapsulates an accumulative and integrated approach to media production and consumption in the early twenty-first century.

"Maximalism" is a term that's been used to describe an aesthetic of digital art, an ethic for technological use, and an ideology of contemporary capitalism, but it similarly encapsulates a common approach to music and media production and consumption. In the context of technology, economics, and creative practice, it serves as an umbrella term for a variety of ideologies related to the mobility and accessibility of media in the digital age. Moreover, it functions as a point of contact for the theory and practice of contemporary music production, highlighting deep-seated anxieties about information overload while serving as a guide for computer music producers as they navigate a range of media platforms.

The maximalist attitude is most noticeable in the general technophobic anxiety among consumers about being unable to keep up with the increasing presence of technology in their everyday lives. In a paradoxical response to the fear that technology may one day render obsolete the agency of the human being, technology is embraced more fully as a means of reasserting individual control over the natural world. Some celebrate the ability to negotiate the constant influx of material technologies and information outlets, while others remain critical of the maximalist claim that technology will solve the world's problems.[42] The perception of media and technology as overwhelming forces in society and culture is certainly nothing new, but the widespread adoption of consumer technology such as smartphones since the first iPhone release in 2007 has made these anxieties even more personal. One iPhone user details the "murky feeling of unease" that arises after witnessing a mother steering her baby stroller with her elbows to free up her hands for smartphone use or upon realizing that he is the only person in a city crowd whose eyes are not glued to the screen of his phone.[43] Media theorist Nicholas Carr examines how web browsing with personal computing technologies has encouraged more distracted forms of human perception and cognition, thus decreasing our ability to focus for extended periods of time.[44]

Maximalism does more than just inspire polarizing debates between Luddites and techno-optimists. It's also a mantra for marketing and productivity in the digital age, a cultural aesthetic, and an economic infrastructure endemic to global capitalism—what anthropologist Frenchy Lunning terms "hyperconsumerism," a bloated economic system that feeds a media market of "monstrous proportions."[45] Theories addressing the increasingly maximal nature of communications media are premised on the notion that globalization has accelerated the rate at which digital products can be produced, accessed, and manipulated, resulting in a constant state of information overload.[46] For media theorist Tiziana Terranova, the consumer should primarily be concerned with how to navigate their way out of the vast sea of digital information and data accumulation in the twenty-first century. As Terranova writes, "network culture" is characterized by "an unprecedented *abundance* of informational output and by an *acceleration* of informational dynamics."[47] Terranova doesn't simply outline changes in consumer dynamics as a result of capitalism at the start of the new millennium, but she also diagnoses "information overload" as an ethical issue for the modern citizen and introduces questions as to "how we might start to think our way through it."[48] Information becomes the material manifestation of an ideology that privileges instant access, technological fluidity, and an omnivorous approach to media consumption.

Paradoxically, the response to this situation by the music products industry has been to produce even more tools for navigating the changing landscape of music software. In line with product marketing strategies across digital media, including downloadable content (DLC) in video games and in-app purchases for mobile apps, DAWs thrive on both stock and third-party plug-ins that provide added functionality for the software. Companies such as Waves and Native Instruments market countless VSTs, audio effects, and sample libraries to music producers endlessly searching for that one kick drum sound or synthesizer preset that will make their composition pop. The "collector's edition" of Native Instruments' *Komplete 13 Ultimate*, for example, includes 115,000 sounds and more than 1,150 GB of content, retailing at a whopping $1,599 (as of March 2021) (figure 1.10). Waves' "Horizon" bundle, which retails at $3,999, includes eighty audio-effect plug-ins alone.

Third-party plug-ins aside, *FL*'s "Plug-in Selector" visual display window allows the producer to choose from 106 stock plug-ins organized by type, ranging from MIDI controllers to effects such as delay, reverb, and filters, as well as sound generators such as samplers or virtual synthesizers, all from inside the *FL* interface. If that's not enough options, the same toolbar includes a shopping cart icon that allows the producer to purchase software upgrades

Figure 1.10 Advertisement for Native Instruments' *Komplete 13 Ultimate Collector's Edition*

Source: Native Instruments, "Komplete 13 Ultimate Collector's Edition," accessed March 23, 2021, https://www.native-instruments.com/en/products/komplete/bundles/komplete-13-ultimate-collectors-edition/).

and extra plug-ins from within the DAW itself, further reiterating music technologist Paul Théberge's idea that making music and consuming technology have become one and the same in the digital age (figure 1.11). Once a plug-in has been selected, it can be dragged and dropped onto the tracks in either the Channel Rack or the Mixer without needing to understand the type or function of the plug-in. Again, this reflects the non-hierarchical layout of *FL* which eschews the traditional workflow and signal flow of other DAWs.

The creation of commodities secondary to the instruments themselves—synthesizer presets, plug-ins, and sample packs that are sold to electronic musicians and producers—has both intensified the male-dominated

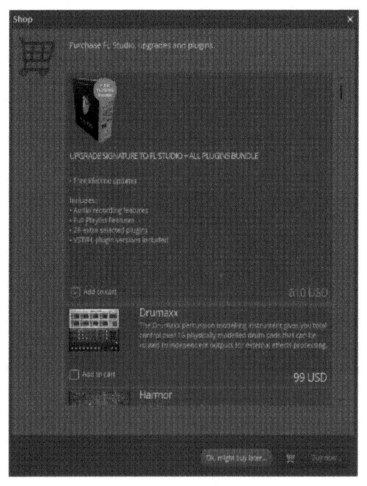

Figure 1.11 *FL Studio* v20 "Shop" (screenshot taken by author).

commodity fetishism and gear lust already prevalent in the music products industry and resulted in the formation of what I call a plug-in culture that has come to define digital music making. The concept of plug-in culture builds off of the masculinist idea of "gear lust," a term commonly embraced within the music products industry to describe consumers of music technology who become addicted to the purchasing of new gear, often going so far as to anthropomorphize their instruments.[49] Whereas gear lust mostly refers to the individual's (often male) commodity fetish for musical instruments, plug-in culture traces this commodity fetish as it becomes institutionalized as ideology within the cultural economy of the music products industry and digital media industries. Plug-in culture functions through a digital maximalist and technological solutionist consumer mentality (presets, plug-ins, apps, and other DLC; the constant upgrading and incorporation of new features in software design, or "feature creep")[50] and a neoliberal form of production in which consumers provide valuable "free" labor for software manufacturers (analytics, user-generated content).[51] Plug-in culture is inherently gendered, tied to a male gaze that analogizes consumption of gear to consumption of women's bodies, a claim that—in light of musicologist Lucie Vágnerová's research on the gendered neocolonial labor of synthesizer manufacturing—goes beyond metaphor.[52] Théberge discusses this in relation to early synthesizer and music technology marketing campaigns that often visualized cyborg women as half woman, half synthesizer.[53] English graphic designer Rachel Laine satirized this element of synthesizer culture when she created a series of nude magazines that equated synth-gear porn with actual porn, crafting headlines such as "Naughty Toys for the Studio" and "A Sawtooth Ripped My Clothes Off." Laine writes in the project description: "I was asked to imagine a series of seventies-era magazines aimed at young men seriously into synthesizers and electronic organs which would all betray their sexist origins by their covers."[54]

Similarly, plug-in culture participates in a long history of cultural appropriation related to the construction of the male neoliberal subject. Think about Diplo, EDM producer-DJ and oft-accused "culture vulture," whose modus operandi involves applying Western, Eurocentric, and Americanized EDM styles to samples from global dance music communities.[55] While Diplo is one of the highest-paid DJs and music producers in the world, most of his "collaborators" from around the world are forgotten once he moves on to another "exotic" location. His technique for crafting patches in Native Instruments' *Massive* synthesizer follows this neocolonial mindset, as he describes sharing presets as global commodities that remove the cultural specificity of global EDM communities: "Me and my crew here at Mad Decent we build and build and build different Massive patches. If you're in the scene you trade Massive patches with other artists, that's what a lot of people do. . . . People love Massive because of its movement."[56] The shared meaning and movement between Diplo's cultural tourism, the internal movement of the digital synthesizer's multiple oscillators and modulators, and the recorded musical commodities produced and globally distributed by the Mad Decent record label all point to a politics of plug-ins in which all parties involved—the human, the machine, the digital audio file—become agents of globalization and neocolonialism. Indeed, the prototypical ubiquitous squelchy lead synth preset made popular by Diplo is a sonic affirmation of DJ Boima Tucker's idea of "global genre accumulation": "No matter where you are in the world, if there's an underground dance scene or marginalized community nearby, Diplo or some DJ like him has or probably will 'discover,' re-frame, and sell it to audiences in another part of the world. 'Western' club DJs are often too stuck in the race for global genre accumulation, to see that the practice of discovery and exposure of Other's culture is always inevitably exploitative."[57] Historically, cultural appropriation was viewed as a type of exoticism. In twenty-first-century culture, where aesthetic terms like "indie" and "underground" lose their meaning in the sea of signifiers that is global popular music, Diplo veils his exoticizing tendencies by describing his "signature sound" as "something that's hopefully eclectic and strange to your ears at first."[58]

In fact, digital audio production has always aligned itself with colonialist paradigms of racial exoticism and conquest. Even more than the global genre accumulation of Diplo and others, the maximalist hyperconsumption of plug-in culture has been fostered by synthesizer and sample-pack companies themselves, as "sounds of local cultures are incorporated into the next generation of electronic musical instruments and redistributed globally by multinational corporations."[59] Think about Native Instruments' "Discovery Series," a collection of sample packs that "capture evocative sound from around the world" and allow musicians to craft their own "world of sounds" using "authentic instruments and patterns."[60] The visual interfaces of the sample packs display in fine detail not only the physical instruments sampled for the sound library but also other "analog" aesthetic markers of "authenticity," including traditional interior decor, such as carpets, curtains, and a general architecture that makes the user believe the sounds were "captured" from the actual music-making sessions of local musicians, rather than being sampled in a professional, controlled recording studio in Los Angeles or Berlin. Electronic musicians and sound designers often spend hours scrolling through the countless presets in these sample packs, giving little more than a passing preview to the sounds included. By claiming to exemplify sonically the experience of Otherness, these presets, plug-ins, and sample packs bring the unacknowledged labor of nonwhite musicians into the capitalist cycle of planned obsolescence, reifying the idea that any sounds (or people) that don't contribute to the final product should be discarded.

The social and cultural effects of this type of software design are clear. Philosopher Robin James sees gear lust and maximalist compositional aesthetics—what she calls "timbrecentrism" and "rhythmic intensification" in EDM production—as examples of how popular music embraces the same techniques, practices, and values of algorithmic intensification and deregulation that neoliberal society uses to manage markets.[61] For James, "new musical technologies take the affective dimensions of musical performance—such as timbral 'sound' and 'feel'—and make them work as one of the central engines of musical composition and expression."[62] In turn, musical value within production communities has come to be defined by the exploitation of a given technology's affordances, as well as gear lust itself—an obsessive search for bigger and better products.

Composer David Bessell calls this relentless pursuit of functionality "feature creep," in which plug-ins and functionality spread from one DAW to another as a result of commercial competition, desire for increased usability, and the need to keep up with technological advances. Examples of this "duplication of function" across music software include the ability to "fix" the timing of recorded audio (quantization), audio time stretching and "warping," tempo detection of audio clips, and video playback from within the DAW, among others (figures 1.12, 1.13, 1.14). ▶[63] As of March 2021, *FL* has undergone 123 version changes since version 1.0 in December 1997, evolving from a simple MIDI sequencer to a full-fledged audio editor and music visualizer. Version 20.6 (2019) alone includes 170 additions, revisions, and bug fixes, including four new plug-ins, demo songs, effect presets, and audio export options, among others. As software developer and electronic music pioneer Miller Puckette notes, "we want software to do everything, and our notion of 'everything' grows broader every year."[64] Daliot claims, "Any company that sells software is engaged in a constant marketing battle with its competitors over features. Typical user behavior is asking why they should

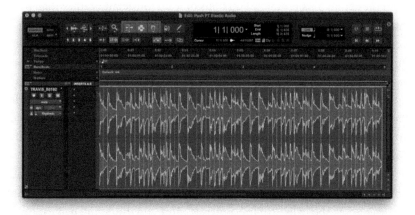

Figure 1.12 "Elastic Audio" in *Pro Tools* v2020 (screenshot taken by author).

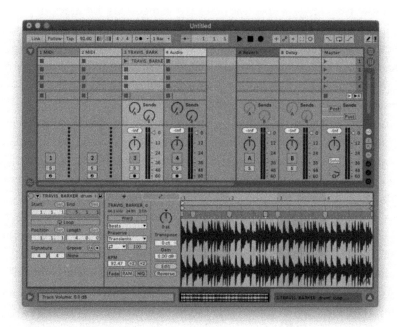

Figure 1.13 "Warping" in Ableton's *Live* 11 (2021) (screenshot taken by author).

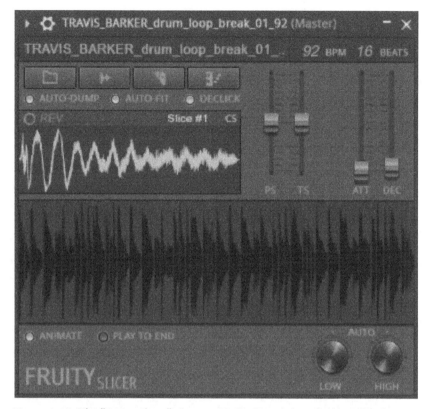

Figure 1.14 The "Fruity Slicer" plug-in, *FL Studio* v20 (screenshot taken by author).

buy your product when your competitor offers twenty more oscillators. . . . There are very few artists that are capable of evading the inflationary over-supply of features. Most people encountering such an instrument act like kids entering a candy shop for the first time. They stuff everything they can get hold of into their mouths and wake up with a bad tummy ache."[65]

Analogizing gear lust and the consumption of music technology to sick-ness and indigestion highlights the very real connections between techniques of music production and maximalist consumption practices under neolib-eral capitalism. James writes: "When you can't expand your market any fur-ther because you're already globalized, the only way to increase profits is by intensifying your current processes, recycling waste into resources. (Naomi Klein calls this disaster capitalism . . .)."[66] Whereas it may have once been perceived as a subversive act to rediscover the cheap and readily available musical resources that society has thrown out, neoliberal capitalism relies on these materials to generate profits in a maximalist and bloated economy.[67] Rather than creating new products or innovative techniques for digital music production, so many music technologies rely on plug-ins, for example, as an intensification of existing technologies and familiar practices.

Musicians have expressed two contrasting responses to the maximalist situation. On one hand, there are producers who are skeptical of the uncritical acquisition and use of new technologies. EDM producer Machinedrum is cautious about the maximalist ethos surrounding the marketing and use of digital tools, echoing Ingram's "crisis of the control surface." He states, "the only problem or 'frustration' I can see coming from music technology would be the seemingly endless amount of options when it comes to software, plug-ins, synths, etc. I feel like a lot of people can become lost trying to get the best gear instead of really honing in on their craft and developing their own sound."[68] The idea of the creative process being stunted as a result of the musician getting lost in a vast sea of studio gear is a symptom of maximalist consumption and reflects a common developmental stage in the introduction of new technologies for music production.

On the other hand, some producers embrace the influx of digital tools as a positive force in shaping the direction of electronic music. Los Angeles–based producer Flying Lotus sees the constant buildup of musical production tools over time as an opportunity for artists to experiment with the limits of electronic music composition and performance: "Why not just have all these things from our past as well as all of the newest technology from today in one, and just really come up with the craziest shit we can? . . . With as much access as we have to all this stuff, to our musical history, our world history, we definitely can be killing shit way crazier. . . . We have the technology!"[69] Significantly, FlyLo conflates access to music technology with a general knowledge of world history, highlighting how technological accessibility is bound up with social and economic values regarding the unfettered acquisition of commodities in consumer capitalism. Here the *accumulation* of technological objects, creative techniques, and forms of knowledge highlights an emerging dialogue between the *aesthetics* of software design and the *technical practices* of composition and performance in digital culture.

Machinedrum and Flying Lotus offer disparate responses to the maximalist situation in contemporary music production. As much as the influx of creative affordances through software can feel liberating and empowering, it can also produce feelings of information overload. Electronic musician

Dennis DeSantis's book *Making Music: 74 Creative Strategies for Electronic Music Producers* encapsulates both the struggle of maneuvering a digital software market that is growing exponentially and the maximalist creative mindset toward composition suggested by DAWs. As Ableton's head of documentation, DeSantis makes a living translating changes in interface design to a dynamic consumer base. However, the book has a different aim. Rather than teaching the technical details of audio production, *Making Music* deals with the psychological hurdles that producers must overcome in order to compose music in the maximalist twenty-first century. Why is music still hard to create in the current "golden age of tools and technology," in which "a ninety-nine-cent smartphone app can give you the functionality of a million-dollar recording studio" and "tutorials for every sound design or music production technique can be found through a Google search"?[70]

DeSantis's project begins from the premise that contemporary life is full of distractions—a logic embedded within the multi-windowed "desktop" environment of the computer itself—and that successful electronic music production is only possible by filtering and limiting these options. In the first section of the book, "Problems of Beginning," he discusses the importance of "resisting gear lust" and "limiting acquisition of plug-ins and virtual instruments" in surpassing the feeling of inaction that often results from confronting the overwhelming nature of digital tools.[71] In line with the modular nature of many DAW interfaces, DeSantis suggests "goal-less exploration . . . the process of simply finding a corner of your working space and letting yourself see what evolves from there" as a nonlinear solution to the feeling that one must incorporate the entire DAW into one's workflow.[72] Focusing on the *process* of learning discrete elements of the software interface rather than simply attempting to churn out marketable *products* prevents musicians from being too constrained by "the *mindset* of these new tools," instead allowing them to concentrate on compositional ideas such as track content, structure, and arrangement.[73]

DeSantis conceives of music composition through sculptural metaphors, emphasizing the process of subtraction in opposition to the additive logic and aesthetic of accumulation inherent in the design and marketing of production technologies. In terms of musical structure, DeSantis suggests starting from the point of "maximal density," organizing the arrangement around that moment, and subsequently chipping away extraneous musical content as needed. Whereas a dominant aesthetic of digital music production views composition in generative terms, the algorithmic nature of digital tools facilitating the seemingly "organic" growth of musical content, chapters such as "Arranging as a Subtractive Process" and "The Power of Erasing" value the *deduction* of content as a strategy in facilitating productivity.

For DeSantis, navigating the maximal interface of DAWs requires a minimalist creative mindset. This paradox of consumer culture is the defining feature of maximalism. As DeSantis's book implies, in order to successfully remove the distractions inherent in technological use in the age of digital convergence, the consumer must keep consuming. In this context of perpetual consumption, maximalism is an expression of an ideology "designed to sell not only a particular commodity but consumption itself," as ethnomusicologist Timothy Taylor suggests.[74]

It's no coincidence that the hypermasculine gear lust, maximalist consumption, and fast-paced production of EDM emerged in the wake of the 2008 economic recession. The fist-pumping, juice-head EDM "bro" was itself a hypermasculist response to a time in which the traditionally male-dominated space of music production was being threatened by plug-ins, presets, and other democratizing forces in music-making software. Archaeologist Peter McAllister shows that in times of perceived crises of masculinity, idealized images of (literally) strong men often emerge.[75] Journalist Pankaj Mishra talks about the "post-9/11 cocksmanship" provoked among many by Osama bin Laden's slurs about American manhood, writing that "one image came to be central to all attempts to recuperate the lost manhood of self and nation: the invincible body, represented in our own age of extremes by steroid-juiced, knobbly musculature."[76] For Mishra, conventional ideals of masculinity are also threatened by economic recessions that often require both men and women to work, thus breaking down traditional household divisions of labor. Industrialization, urbanization, and mechanization always, in turn, reify strict gender norms and perpetuate the myth of the selfmade man on a "futile hunt for individual power and wealth." The story of *FL* is the story of the music products industry itself: a space in which the increasingly threatened gendered division of labor in music production has resulted in a hypermasculine gear lust for bigger and better tools. In the 2010s, especially, the DAW became a technology of resilience for white male EDM producers looking to pursue the American dream, materialized as a mythos in the 1 percent of DJs who could land a Vegas residency or a spot on the *Forbes* list.

2

Monopolies of Competence

In 1990, Palo Alto–based software company Opcode released *Studio Vision*, the first software to integrate audio editing with MIDI notation, merging the previously disparate tools of musical instruments and personal computers. Sporting a simple GUI that arranged both audio waveforms and digital music notation along a linear timeline, *Studio Vision* responded to a popular desire among musicians looking to compose recorded music using nothing but their personal computers and a $995 software program (figure 2.1).[1] While *Studio Vision* established an enduring design layout for contemporary DAWs, the simplicity and purposeful functionality of the software is a far cry from the overwhelming complexity that would emerge in the design of DAWs just two decades later.

In the twenty years following the release of *Studio Vision*, countless DAWs emerged, each software expanding on the functionality of previous programs. Some, such as Reason Studios' *Reason*, modeled themselves on the design of "analog" hardware devices prevalent in physical recording studios, allowing users to incorporate hundreds of synthesizers, effects boxes, and sampling machines into their production setups. Others, such as Ableton's *Live*, *Bitwig Studio*, and *Renoise*, arranged their GUIs as vertical grids in which musicians could endlessly stack musical patterns on top of one another. As a result of the ever-increasing creative options available in the age of software, musicians have learned constantly to adapt existing compositional techniques to new interfaces. In turn, DAWs have become more than just functional tools for the purposes of audio recording and composition; they've become experimental playgrounds and media environments that extend beyond the physical confines of the "analog" music studio. ⏵

Throughout the history of DAWs, and the history of music technology more broadly, marketing tropes have focused on the ways in which the tools are increasingly more accessible and easier to use, therefore "democratizing" the practices of music making.[2] Yet the rise of social justice movements such as Me Too in 2017 revealed the dearth of women in professional roles across the creative industries, as well as the forms of sexual harassment taking place

Push. Mike D'Errico, Oxford University Press. © Oxford University Press 2022.
DOI: 10.1093/oso/9780190943301.003.0003

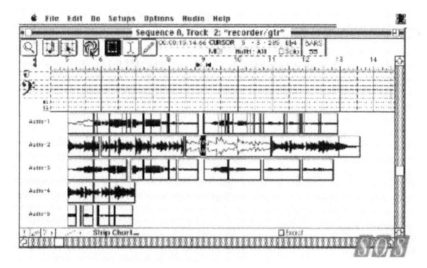

Figure 2.1 Opcode's *Studio Vision* (1990)

Source: "25 Products That Changed Recording," *Sound on Sound*, November 2010, accessed August 27, 2020, https://www.soundonsound.com/reviews/25-products-changed-recording.

in and out of professional settings that have contributed to this lack of representation. As a result, the music products industry began altering its marketing campaigns, introducing popular feminism as a neoliberal strategy of aligning product brands with the personal identities and experiences of a diverse consumer base.

Is this not simply another phase in the long history of capitalist marketing campaigns in which music technology is uncritically praised for its transformative, empowering, and democratizing potential without effecting actual changes in equity?[3] Education research continues to prove the massive gender differential in girls' and boys' computer use in the classroom, much of which has been attributed to socialized expectations of male and female attitudes toward computing.[4] This chapter asks two seemingly unrelated questions. First, how do changing trends in musical instrument and interface design shape techniques of electronic music composition and performance? Second, how might the specific design affordances and constraints of DAWs such as *Live* and *Pro Tools* reflect the social dynamics of gender and identity at the heart of the software and entertainment industries in the 2010s? After comparing approaches to the design and use of *Live* and *Pro Tools*, I tease out how these software programs reflect the values and ideologies of the communities surrounding them to think through ways of fostering actual changes to diversity, equity, and inclusion in the music software industry.

Generally, DAWs follow two different design approaches that shape their GUIs, workflows, and core affordances and constraints: instrumental and computational. In what I call the *instrumental* model of software design, the relationship between "new" and "old" technologies is isomorphic, as the emerging tools seek to mimic the look and feel of existing tools. This tool-based notion of instrumentality aligns with philosopher Martin Heidegger's first definition of technology as a means to an end.[5] Following this definition, a commonly understood function of software is to make more efficient the human tasks previously completed with hardware. For example, it could be argued that the reason for the success of iPads and tablets is that they resemble notepads or that the use of metaphors such as "desktops" and "folders" has allowed the personal computer to thrive as an office tool. Media theorists Jay Bolter and Richard Grusin define this relationship as "remediation," in which a new media platform "appropriates the techniques, forms, and social significance of other media and attempts to rival or refashion them in the name of the real."[6] Designers refer to this use of interface metaphors as skeuomorphism. As literary critic Katherine Hayles writes, "*skeuomorph* is a term anthropologists use for a device that once had a functional purpose but in a successor artifact loses its functionality and is retained as a design motif or decorative element."[7] Music technologist Adam Patrick Bell notes that "a great advantage of software incorporating skeuomorphism is that new users are provided with visual cues of familiarity, easing the difficulty of making the transition from hardware to software."[8] From this perspective, the success of new interfaces is dependent on the extent to which the fundamental logic and cultural understandings of previous tools and techniques are retained in their software simulations. However, Bell also points out the "significant downside to skeuomorphic design philosophy, namely, that new technologies are used for old ways of thinking and doing."[9]

Historically, *musical* instrumentality has been defined in a similarly tool-based manner. As a discipline, organology—the study of musical instruments—arose from a functional definition of instruments that includes both the sounds they produce and how the instruments produce those sounds.[10] For ethnomusicologists concerned with the historical preservation of "traditional" musical instruments, instruments are often perceived to be "fixed, static objects that cannot grow or adapt in themselves" and therefore can only exist as tools for the purpose of human expression and creativity.[11]

Musician Aden Evens echoes Heidegger's definition of technology in his own description of instrumentality, writing that when people "refer to a tool, a technology, or a person as an instrument, part of what we intend is a reduction of that tool, technology, or person to its instrumentality. That is, an instrument is something that serves a particular end, and, as instrument, it is merely a means to that end."[12] Instrument *design* generally aligns with this tool-based, functional understanding of instrumentality.[13] As with any technological development—musical or otherwise—new instruments are often designed to retain recognizable elements from previous instruments, while offering an attractive enough set of new affordances to allow the tool to be widely used.[14] In the case of electronic musical instruments, "traditional" instruments are often simulated by the computer, as virtual *tools* for the purpose of recording musical *content*.

Software such as Avid's *Pro Tools*, Apple's *Logic Pro X*, and Reason Studios' *Reason* remain the most common examples of these types of tools. DAWs facilitate the instrumental model of software design in three ways: first, through the use of add-on plug-ins to emulate existing musical tools; second, through the incorporation of a horizontal timeline recording layout to emulate the linear temporality afforded by previous musical media such as vinyl records and magnetic tape; and third, through the general use of skeuomorphs that resemble physical objects. Together, these features retain the assumption of music software as an object-oriented tool designed for the purpose of audio inscription and recording.

As mentioned in chapter 1, VST plug-ins are software add-ons that provide additional functionality when used alongside a DAW. Plug-ins exemplify the instrumental model of software design in that they're often designed as virtual metaphors for existing physical instruments. That is, in order to negotiate the transition from hardware devices to software programs, "digital" plug-ins imitate the look and feel of musical tools that currently only exist in "analog" form. A survey of popular plug-ins includes software *emulators* of traditional instruments (figure 2.2), popular drum machines such as the Roland TR-808, and a vast array of vintage synthesizers (figure 2.3), as well as *simulators* of common techniques employed by producers while using these tools, such as grid-based rhythm and note sequencing, compression (figure 2.4), basic sound synthesis, and audio sampling. ▶

In addition to the incorporation of plug-ins, the instrumental design approach in DAWs is fostered by a horizontal timeline recording layout that emulates the linear temporality afforded by previous media forms. In *Pro*

Figure 2.2 AIR *Mini Grand* plug-in (screenshot taken by author).

Figure 2.3 Korg *Lexington* plug-in (ARP "Odyssey" emulator) (screenshot taken by author).

Figure 2.4 Native Instruments *VC 2A* compressor/limiter plug-in (screenshot taken by author).

Figure 2.5 *Pro Tools* 12's "Edit Window" (screenshot taken by author).

Tools' "Edit" window (figure 2.5) and *Logic Pro X*'s "Timeline" (figure 2.6), for example, the GUI is organized as a horizontal, linear recording timeline meant for the layering of multiple tracks. Producers are already familiar with the process of organizing musical ideas into complete, precomposed structures—a fact that's evidenced by the ever-present linear timeline across DAWs. By remediating the physical processes and graphical layout involved in *writing* musical ideas on sheet music or *splicing* and rearranging magnetic tape, the timeline provides a visual analogy to earlier, analog forms of musical composition.

Figure 2.6 *Logic Pro X*'s "Timeline" (screenshot taken by author).

Further, DAWs foster the instrumental design model with design metaphors that resemble analog interface elements such as sliders and knobs. Since the advent of personal computing in the 1980s, these types of index-ical icons and material metaphors (skeuomorphs) have guided the design of computer operating systems.[15] For example, the "Mailbox" icon analogizes email to snail mail; the "Trash" icon aligns the practices of discarding dig-ital files and throwing away physical waste; and "Folders" on the computer "Desktop" are meant to store digital files in a similar fashion to the manila folders of an actual desktop. In each example, indexical icons are meant to relate tasks in the digital software environment to everyday tasks with which the computer user is already familiar. In this way, the suggested usage of the software is implied by the design metaphor, thus requiring little to no spe-cialized knowledge from the user.

Material metaphors in the form of buttons, knobs, and sliders are present throughout the GUIs of *Pro Tools*, *Logic Pro X*, and *Reason*, among others. When adding effects to a given track in *Logic Pro X*, the musician can ma-nipulate specific effect parameters using minimalist interface elements that abstractly represent the rotary knobs on an analog synthesizer (figure 2.7). Similarly, adjusting the volume of an audio track in *Pro Tools* requires the manipulation of a vertical slider that represents a simplified version of the faders that appear on physical mixing boards. In addition, muting, recording,

Figure 2.7 *Logic Pro X* compressor plug-in (screenshot taken by author).

Figure 2.8 Channel strip from *Pro Tools* 2020's "Mix Window" (screenshot taken by author).

and soloing individual audio tracks can be achieved by clicking one of three rectangular buttons, thus mirroring the design and functionality of a mixing board (figure 2.8). *Reason* incorporates an entire rack of 3D hardware effects, complete with screws and bouncy patch cables that simulate real-world physics (figure 2.9). In these examples, the skeuomorphic design of the interface metaphors serves to focus the musician's attention on the physical task required of each element, whether that be sliding, turning, pressing, or switching. In other words, design metaphors encourage the computer users to forget that they're interacting with software at all, instead focusing their attention on the physical hardware simulated by the GUI.

As the examples of plug-ins, timelines, and other interface metaphors demonstrate, remediation is a common factor in digital musical training. These tools are designed based on the philosophy that for musicians to develop technical proficiencies on new instruments, interface design should ideally resemble previous instruments in some way. However, in translating analog technologies to digital software, instrumental design models

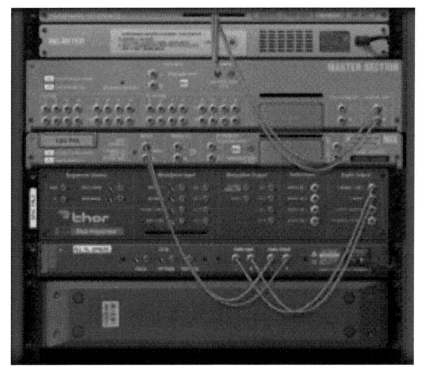

Figure 2.9 *Reason* 11 "Rack"
Source: https://www.reasonstudios.com/en/reason/rack-plugin.

risk reifying a limited set of practices and stunting the creative possibilities of musicians. Musicologists have critiqued how musical media such as sheet music and recordings encourage relating to the music as a fixed text or "work."[16] Similarly, instrumental design models privilege an understanding of software as a *text-based* medium for the purpose of inscribing existing objects and practices. More important, instrumental design—in favoring GUIs that emulate existing technical and compositional practices—risk alienating demographics of users who have never had access to musical technologies such as recording studios and robust computer workstations or the privilege to study music and pay for instrumental lessons throughout their lives. Design researcher Ann Light writes about how not only does working with design "norms" marginalize those outside the "normal" group, but "in reinforcing cultural norms through encoding them in our interfaces and underlying systems without room for alternatives, we have also made a contribution to our future: certain paths are less obvious or less possible."[17] Outside of financial and class accessibility, instrumental design can covertly foster forms of gender discrimination that lead to social alienation in music production communities.

Visual elements of design don't simply define the creative possibilities of a software application. They also inform what design theorists Lars-Erik Janlert and Erik Stolterman call the "character" of software. In what they call "character modelling," people attribute characters to artifacts to handle complex information: "rounded forms and warm colors suggest that the car (or whatever) has a warm, friendly, and protective character. We tend to assume a connection between the mere appearance of the artifact and its character. . . . The rationale is that physical appearance sometimes reveals important information about inner structure and the way the artifact will behave."[18] Architecture and industrial design employ similar product semantics and character signaling to guide users through the space; architects refer to the underlying character of a space as its "parti."[19] The main question, then, is not simply how design affordances and constraints shape creative practices but also how approaches to design reflect a social character in the software that is, in turn, reflected in the community of users surrounding the program.

Pro Tools reflects an instrumental design approach that perpetuates what I call monopolies of competence[20]—that is, the use of skeuomorphic design in "professional" applications to perpetuate hierarchies of users, in contrast to overtly user-friendly design that may appeal to a wider demographic of users. The affordances of *Pro Tools*—including its skeuomorphic design rooted in studio practices from a predigital era of recording—cater to professional users working in the increasingly exclusive space of the recording studio, and the user base embraces these specialized skill sets as signs of their own professional competence. This mindset is encouraged by the titles of software such as *Pro Tools* and other DAWs that have historically relied on skeuomorphic design such as Apple's *Logic Pro X* and Reason Studios' *Reason*. Even the term "digital audio *workstation*" implies a tool used by professionals for business purposes rather than personal use.

"Whether or not design-speak sets out to colonize human activity," sociologist Ruha Benjamin writes, "it is enacting a monopoly over creative thought and praxis."[21] How exactly are monopolies of competence reflected in the design of *Pro Tools*? Discourses of an outmoded professionalism are evident in some of the software's design failures, including the use of skeuomorphs that no longer exist or that are no longer possible given the nature of computer operating systems in the early twenty-first century. The lack of certain features creates barriers of entry to the software, requiring users to purchase extra hardware in order to use the software to its full capabilities. For example, it wasn't until 2021 that an in-app typing functionality was implemented, meaning that for thirty years, the use of a MIDI keyboard was required to sequence MIDI instruments in real time with *Pro Tools*. Also, it's very difficult to fit the information of both the "Mix" and "Edit" windows into a single screen, so most users work with two display monitors. Unlike other DAWs, the "Mix" window isn't dockable, which further prevents single-screen usage.[22] Other issues include long-term problems with screen resolution when using newer display formats such as 4K or retina. Finally, a small but important issue brought up by many *Pro Tools* users is that you can't undo the deletion of a track. If we're thinking back to the "character" or "parti" (to follow the architectural concept) of *Pro Tools*, common descriptors might include—and I've certainly heard many users describe the character of the software in this way—stubborn, untrustworthy, self-confident, conceited, demanding, and bossy.[23]

Professionalism is further evoked by material and tangible signifiers that one might find in a professional recording studio. Architect and designer Matthew Frederick writes about how soft implements are valuable tools for exploring conceptual ideas early in the design process, whereas hard-line drawings are best for conveying information that is decisive, specific, and quantitative, such as final floor plans.[24] In addition to the 3D textures and perceived tangibility of *Pro Tools*' skeuomorphs, the contour of the software's windows uses hard edges and representational graphic elements that connote decisiveness, specificity, and finality. Pop-up windows like the "Transport" window (figure 2.10), plug-in windows (figure 2.11), and the "Workspace Browser" (figure 2.12) use beveled and shaded buttons with sharp edges, as well as 3D skeuomorphs like rotary knobs and folders to convey materiality. This design approach imbues a sense of stability in the overall aesthetic that mirrors the standardization that many professional users strive for in order to optimize their creative workflow.

Figure 2.10 *Pro Tools* 12 "Transport" window (screenshot taken by author).

Figure 2.11 *Pro Tools* "DVerb" plug-in (screenshot taken by author).

Figure 2.12 *Pro Tools* 12 "Workspace Browser" (screenshot taken by author).

Finally, monopolies of competence are perpetuated by the sheer number of options and interface elements available in *Pro Tools*, echoing the maximalist aesthetics presented in chapter 1. *Pro Tools* has been around for a long time, and it's had to adapt to increasingly changing industry needs, but the designers have mostly approached updates from an additive perspective, leaving old features and GUI elements intact while slowly introducing new elements. This additive approach to software development is typified by the subscription model that's become highly popular among software companies, including Avid. In this model, users pay a monthly fee for the software, rather than a one-time perpetual license. The motivation is that users will receive constant updates with new features every month; the more features, the better. The result is a lack of framing, a busy agglomeration of interface elements that only make sense intuitively to legacy users, decreasing the software's appeal to new users without professional training or traditional expertise in audio production. Echoing Frederick once again, "the more specific a design idea is, the greater its appeal is likely to be . . . designing in idea-specific ways will not limit the ways in which people use and understand your buildings; it will give them license to bring their own interpretations and idiosyncrasies to them."[25] The type of "feature creep" exemplified in *Pro Tools* has the effect of limiting broader user appeal by confusing the main idea of the interface's design aesthetic.

From a UX perspective, there's nothing wrong with designing a project for a specific user base of industry professionals. However, when that user base is consistently defined as made up of a homogeneous racial and gender demographic while the "professional" status of the software is challenged by countless other consumer-level DAWs, battles over hierarchies of taste and professionalism in audio production circles become battles over who gets to participate and who gets excluded. Music education expert Victoria Armstrong talks about how music-making software presumed to be "difficult" is perceived as "masculine," whereas software experienced as "easy" tends to be characterized as "feminine."[26] Skimming any given pro audio web forum will immediately make apparent how monopolies of competence are used to establish expertise and reify long-held gendered divisions of labor in music production communities. Typical conversations include anecdotes about *Pro Tools* users as "snobs" who think their DAW is superior to tools like *Live* and *FL Studio*, which are often denigrated as "toys."[27] For decades, the idea of *Pro Tools* as the industry standard has allowed it to carry a perceived gravitas as a "pro" DAW against all the rest. Nowadays, the notion that there even is a standard DAW is being tossed out the window, since so many professional productions are being created in smaller, project studios rather than the big-budget studios that might require a larger *Pro Tools*–based setup.[28]

Musical practices associated with power and leadership, such as conducting, composition, or audio engineering, have long been stereotyped as masculine domains. As practices that require knowledge and control of technology and technique, these skill sets have been historically and socially constructed as masculine displays of expertise, or what Lucy Green calls "metaphorical display(s) of the mind."[29] This manifests itself primarily through perceptions of the music technologies themselves, including the perceived "masculinity" of DJ and music production software, as musicologist Tami Gadir notes.[30] It's no coincidence that software such as *Pro Tools*, *Reason*, and *Logic* are so often linked to masculinity, since masculinity is itself often defined by attempts to assert monopoly over the control of reason, logic, and objectivity. Satirizing this point, music writer and comedian Kyle Vorbach wrote an article for *The Hard Times* titled "Ben Shapiro Produces EDM EP Using Logic and Reason," in which the right-wing political commentator claims: "Digital audio workstations don't care about your feelings. Apple's Logic Pro X and Propellerhead's Reason were the only logical, reasonable choices for my new EDM project. . . . Let's look at the facts here: the free market bolsters creativity and competition. And with its innovative rack interface and powerful sequencing features, so does Reason."[31] In the spirit of the best satire, the piece leaves us with the perennial question of whether this is clever comedy or sad reality. Discourses of professionalism in the United States are agents of white supremacy and hegemonic masculinity, functioning as gatekeepers that decide who gets to participate and who gets left behind. As Bell notes, "there are parallels between the dominance of whiteness and masculinity as they relate to the practice of playing the studio . . . the crux of this imbalance on both fronts is the hegemony of white male recording culture."[32] If masculinity is often framed in terms of technical skill and expertise, the monopolies of competence fostered in the design and use of *Pro Tools* alienate a more diverse user base.[33]

Nowhere are these monopolies of competence more evident than in music technology education, where boys often flaunt their knowledge as a way of demonstrating mastery over the subject. Musicologists Georgina Born and Kyle Devine discuss the "indirect discrimination" that occurs in the music technology classroom, including the use of a "discrete critical vocabulary" surrounding music production terminology and a compositional style that distinguishes "male" as virile and powerful and "female" as delicate and sensitive.[34] On a more fundamental level, Armstrong describes how tech-oriented classrooms and other creative spaces are often "discursively, atmospherically, and spatially male-dominated."[35] Following the discussion from chapter 1, we might also consider the "gendered social imaginaries" that take shape surrounding specific instruments and how those connotations create unequal gender relations.[36]

To understand the social dynamics of software use, one simply has to imagine the ideal user for a given program. In the context of *Pro Tools*, the ideal operator is a "professional" who works above the everyday "user" and therefore doesn't perceive the need for the type of guidance afforded by principles of UX and interaction design. Learning the seemingly complex "language" of the software becomes a barrier of entry and a rite of passage for the aspiring producer. This is no different from long-standing debates about who gets to count as an artist (in the visual arts) or a composer (in the musical arts), terms generally assigned to those who have mastered the training of the Western art canon. In the Western art music tradition, composers and professionals garner the respect of being able to produce "works," as opposed to popular musicians who make beats or write songs. This distinction between valued labor (work) and leisure (play) defines a core difference between instrumental and computational design approaches. The concept of the musical "work" as an *object* may have made sense in the context of music printing and vinyl record distribution, but the *process-oriented* nature of software equally foregrounds the importance of algorithms, procedures, and rule-based systems.

As much as the tool-based nature of *Pro Tools* appeals to computer musicians with some traditional training in audio production or musical performance, other software—such as Ableton's *Live* and *Bitwig Studio*—incorporates computational affordances that depart from representational and metaphorical design trends. Undoing the monopolies of competence in music software design requires a critical design practice that starts by challenging the "normal" trends that have gone unquestioned for so long. Design researcher Shaowen Bardzell writes that "researchers, designers, industry product strategists, and others in privileged positions of power should not merely be experts in a given domain or technology but must also have enlarged mentalities—the discipline to confront the otherness of the other and to change and grow as a result of it."[37] How does Ableton incorporate computational design trends to construct and teach alternative users of

DAWs? More important, are changes in software design enough to promote actual diversity, equity, and inclusion in the music products industry?

In 2001, the Berlin- and Los Angeles–based software company Ableton released *Live*, the now standard DAW for music recording and performance. Until the time of *Live*'s release, DAWs catered mostly to demographics of amateur and professional musicians interested in how software could make their *recording* and playback setup more efficient. In contrast, *Live* was designed for both recording and, as its name suggests, "live" *performance.* Considering this goal, the GUI of the software was designed in such a way as to facilitate rapid audio editing, the ability to loop digital samples with ease, and the seamless integration between tools for recording and performance (figure 2.13). The buttons, knobs, and sliders of "analog" mixing boards are

Figure 2.13 Ableton *Live* version 1.5 (2002)

Source: "Ableton Live Audio & Loop Sequencing Software for Mac and Windows," *Beat Mode*, n.d., accessed April 25, 2016, http://www.beatmode.com/historical/ableton-live/.

simulated in *Live*, alongside features unique to "digital" software: a vertical grid for arranging digital samples in a nonlinear fashion and an array of windows for editing audio at a micro level. Significantly, *Live* also included a drag-and-drop interface that allows for the real-time remixing of musical content.

By integrating existing compositional techniques with nontraditional forms of musical interaction, software such as *Live* represents a twofold challenge for the designer, as the digital musician is asked both to navigate a foreign set of performance techniques and to embody a non-"musical" physical interface to the computer (the mouse and keyboard do not control a fixed set of musical parameters but can also be used to check emails, draft manuscripts, and play video games). Historically, the design and development of music software has strived to strike a balance between the simulation of existing musical interfaces and techniques on the one hand and the introduction of new creative possibilities on the other. *Live* reflects, on one hand, an *instrumental* lineage that bases the design of new interfaces on pre-existing musical instruments and software. On the other hand, *Live* reflects a *computational* lineage that introduces process-oriented mechanics and "flat" design into music software. Through analyses of the GUI elements in *Live*, we can see how both contemporary forms of music composition and software design are rooted in iterative, modular, and abstract *processes* rather than fixed *objects*. Taking this into consideration, I align digital musical composition and performance with the procedural nature of software code: as a set of process-oriented endeavors which require constant "updating" through new forms of musical training.[38]

Software is built on computational affordances that either extend or depart from representational and metaphorical design trends. While plug-ins *represent* existing musical tools and techniques, DAWs also offer the unique capability of *abstracting* what might be considered more traditional "musical" tools and techniques, thus presenting new possibilities for digital composition. In contrast to the instrumental design model, what I call the *computational* model of software design embraces the unique affordances of the computer as both an instrumental medium for the authoring of text, audio, and visual content and a process-oriented medium for the authoring of processes themselves. If *tools* such as word processors and photo editors exemplify the instrumental model, the computational model is best represented by rule-based *practices* such as programming code, generating algorithmic scripts, and designing digital games. DAWs such as *Live* facilitate the

computational model of software design in three ways: first, the incorporation of a nonlinear temporal structure in the form of a vertical grid; second,
the use of "real-time" feedback mechanisms that afford the rapid and iterative prototyping of musical material; and third, the ability to drag and drop
interface elements in a modular manner. Together, these GUI elements present a non-skeuomorphic design approach that contrasts with the monopolies of competence encouraged by *Pro Tools.*

In contrast to the horizontal timeline of the *Pro Tools* "Edit" window and
Live's own "Arrangement View," *Live's* "Session View" organizes musical
patterns into a vertical grid designed to facilitate the real-time rearrange-
ment of musical "clips." In this *computational* design model, musical patterns
are organized as literal building blocks of data that can contain anything
from a short vocal sample to an entire multimovement symphony (figure
2.14). "Session View" encourages a shifting understanding of the musical
"work" by fragmenting the compositional process into modular units. This
nonlinear aspect of the GUI remains the most radical and appealing feature
of the software for many producers. Beats By Girlz founder Erin Barra claims

Figure 2.14 Ableton *Live* "Session View" (screenshot taken by author).

that "the product itself, Ableton Live, is non-linear and Ableton is largely the only non-linear DAW on the market. As a musician, very infrequently are we experiencing music from beginning to end. I gravitate towards that tool because it mimics the way that people actually compose for practice or produce. It's a real reflection of the music making process."[39] DJ Mike Huckaby claims *Live* to be "the most revolutionary sequencing and production software to emerge from the past 10 years," largely due to how the clip grid "destroyed the concept of linear thinking while making music."[40] By providing an alternative to the horizontal timeline, *Live*'s sketchpad-like temporal structure fragments the notion of the holistic, self-contained musical work and intensifies the process-oriented nature of composing with digital software. ⏵

In addition to the use of a modular temporal structure, *Live*'s computational design is reflected in the incorporation of "real-time" feedback mechanisms that allow for rapid and iterative prototyping of musical material. As the name suggests, *Live* signifies mobility, emphasizing the creative flexibility that results as users transport the software using their portable laptops. From a practical perspective, the seamless workflow in which the GUI immediately displays the results of various actions by the producer (figure 2.15) also adds to the real-time sensibility. Composer and multi-instrumentalist Angélica Negrón discusses how she uses *Live* to audition and preview her work while writing it, allowing her "to listen to both elements (electronic and acoustic) simultaneously, which is something I'm not able

Figure 2.15 *Live* 9 promotional material
Source: Ableton, "Live," accessed April 22, 2015, https://www.ableton.com/en/live/.

to do in notation software."[41] Negrón describes a composition process that exemplifies iterative design methodologies. That is, *Live* affords Negrón a process in which the creative materials are constantly reworked based on user testing and real-time evaluation of sonic materials, something she's unable to carry out in more traditional notation software.

In contrast to the typical recording studio model which values the *inscription* of musical material onto physical storage media, the real-time aspect of composing in *Live* instead privileges the fine-tuned *manipulation* and *reperformance* of material. Studio production blurs with stage DJing, as the producer pulls sonic material from their "Library," layering track after track without interrupting the workflow. As the music runs, the most minute micro-parameters of the audio signal can be finely edited, with each virtual turn of a knob capable of being automated and recorded within each clip. For example, to create a four-bar looping drum pattern that fades out over the course of each iteration, the producer literally draws a downward "automation" curve over the waveform of the pattern (figure 2.16). In this way, *Live* encourages not only the reperformance of existing musical material but also the inscription and automated playback of performance gestures such as knob turning, slider fading, and button pressing. This ongoing, iterative cycle of production and reproduction prioritizes the performative aspects of sound *manipulation* rather than the inscriptional aspects of sound *recording*. In doing so, real-time feedback mechanisms foreground a shift in the conception of software from text-based code (object) to "live" instrument (process). ▶

The third defining feature of *Live*'s computational design is the modularity of interface elements. Whereas the instrumental model of design values faithful recreations of existing musical instruments, the layout of the *Live* GUI can be redesigned and reassembled by each individual musician. This malleability is most clearly present in one of the primary design functions of

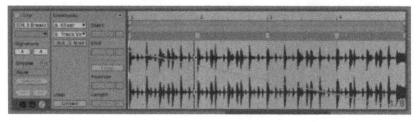

Figure 2.16 Automation curves in *Live* 10 (screenshot taken by author).

Live: the ability to drag and drop nearly every musical element into another. Sound effects can be dropped onto individual musical clips, entire tracks can be dropped into the middle of an extended musical arrangement, and media from outside the *Live* set can be dragged and dropped into the current workflow (figure 2.17).

Significantly, the drag-and-drop function can be performed without disrupting the audio currently being played, thus allowing for the further manipulation of musical events in real time. Experimental electronic musician Synnack describes how this functionality has allowed him to overcome the "nightmare" of exporting and importing musical material into a coherent workflow: "The fact that, in *Live*, you can just drag and drop songs into each other, and even preview them from the browser was totally revolutionary and still is. . . . The fact that you can drag and drop effects into a song while it's playing, and even reorder them with no audio dropouts is insane."[42] For Synnack, the primary benefit of drag-and-drop in *Live* is that it does not disrupt the playback of the musical tracks, thus encouraging iterative

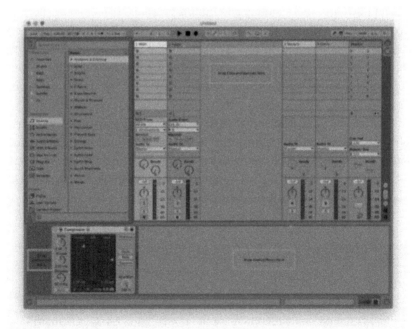

Figure 2.17 Drag-and-drop functionality in *Live* "Session View" (screenshot taken by author).

prototyping of the composition. Live electronic music virtuosos such as Sowall,[43] Laura Escudé,[44] and Sakura Tsuruta[45] use the Ableton "Push" MIDI controller to make the drag-and-drop workflow even more quick, efficient, and tangible. Likewise, Suzi Analogue claims, "I'm really guided by what gear I have around. A lot of the record is me exploring the capabilities of the Push 2, getting to know it."[46]

Drag-and-drop affords the modularity not only of musical material but also of the interface elements themselves. Like most DAWs, *Live* incorporates a traditional windows, icons, menus, pointer (WIMP) interface that makes transparent the various screens through which the producer composes. Additionally, in *Live*, each segment of the production workflow—the user's sound library, the transport (play, stop, record) controls, sound effects, and track content—appears as a distinct module, clearly delineated by a thick, rounded frame that functions as a border to the corresponding window. As mentioned earlier, soft borders in frames are employed by architects as a tool for exploring conceptual ideas early in the design process. Media theorist Steven Johnson singles out the development of the design feature "windows" as the most important interface innovation in personal computing, as it facilitates a "more layered and multiplicitous" onscreen space that allows users to quickly switch between various modes of thinking and practice with the click of a mouse.[47] In the context of the *Live* GUI, the WIMP interface with a softer frame encourages mobility, accessibility, and modularity of not only the musical *content* of a given production but also the *form* of the DAW itself.

Existing scholarship on the relationship between musicians and technology has focused on how musicians use technologies designed for non-musical purposes as musical instruments. Mark Katz and other musicologists describe the virtuosic performances of hip-hop DJs, specifically aligning the creative practices of turntablism with those of classical musicians.[48] David Bernstein examines the use of tape recorders and other multimedia tools as performance instruments in the 1960s avant-garde.[49] In each case, technologies not intended for music creation are valued for their capacities to become tools for the purpose of musical performance. In contrast, the *Live* GUI exemplifies a convergence of instrumental and computational design models. As such, *Live* is more than just a musical instrument or tool for sound recording; it's an exploratory environment of creative affordances through which the musician navigates. Examining *Live* as a process-oriented environment rather than a fixed instrument allows us to shift the focus from the GUI itself to the broader network of tools and techniques that continue to shape the design and use of the software.

Within technology circles, the term "environment" is often used to refer to the external or internal factors that shape the design and use of a given technology. The "software development environment" refers to the programming tools used to create the software. The scholarly discipline known as "media ecology" specifically studies media and technology as environments that structure human perception, feeling, and value. Environmental metaphors are commonplace among professional interface designers. Borrowing a term from information science, for example, we might think about how the *Live* GUI epitomizes an open and transparent "information architecture."[50] Initially defined by architect and graphic designer Richard Saul Wurman as the creation of "systemic, structural, and orderly principles to make something work—the thoughtful making of either artifact, or idea, or policy that informs because it is clear,"[51] information architecture now encapsulates various meanings and practices across information science and technology. These include the structural design of shared information environments,[52] the art and science of labeling websites and other information management programs,[53] and a community of practice focused on migrating design principles to the digital landscape.[54] *Live*'s computational design approach can be understood as an attempt to construct a broader "environment" in terms of both the technical affordances that frame the digital musicians' creative workflow and the demographics of producers who use the software.

Analyzing *Live* through the structural metaphor of information architecture provides a model for understanding the DAW as a software environment above and beyond an instrument. In physical recording studios, glass windows often serve as barriers between performance spaces in which musical content is recorded and control rooms in which sound is managed and edited during and after the performance. When this design principle is translated into digital software, physical space is flattened into a single screen, removing the barrier between the creative spaces of the stage (as a performance space) and the studio (as a sound management and editing space). In removing the *glass* that fractures and isolates the production process into separate spaces, sound in the digital environment can filter and bleed through the various elements of the computer *screen*.

Most immediately, the architectural metaphor may evoke an instrumental conception of software as object—as "frozen music," perhaps. Magnusson applies this concept to the cultural politics of software, as he writes about how "values, opinions, and rhetoric are frozen into codes, electronic thresholds, and computer applications. Extending Marx, then, we can say that in many ways software is frozen organizational and policy discourse."[55] Following this Goethean imagery, I suggest that the process-oriented nature of composing in *Live* exemplifies software as "liquid architecture": an environment structurally defined by the movement, juxtaposition, and recombination of sound and interface elements, rather than through fixed visual structures. "All that is solid seems to melt in the cloud," as media theorist Marianne Van Den Boomen puts it.[56] The modularity of musical ideas occurs first and foremost within the open framing of the GUI (e.g. the WIMP interface, the ability to drag and drop musical tracks, and the use of interface metaphors), guiding producers as they navigate the interconnected structures of the various interface elements.

At the same time, framing the DAW as environment also captures how the software extends *beyond* the centralized space of the *Live* interface, as an interoperable program within a broader system of digital tools and techniques. Musicologists Jason Stanyek and Benjamin Piekut describe "leakage effects" as inherent qualities of digital recording spaces, occurring "when an activity in one area expands unexpectedly into another area, setting in motion a second process, project, or concern."[57] For Stanyek and Piekut, the productive potential of software manifests in the ongoing convergence of music composition tools and techniques across media. These leakage effects occur not only when musical content (plug-ins, audio files, and effects) moves between the various modules of *Live*'s sonic architecture but also when the previously fixed space of the digital "studio" extends beyond a single dedicated software. This software convergence forms what Théberge calls a "network studio," influencing the techniques and practices of other applications within the computational environment.[58] The metaphorical perforations inherent in the screens of software thus deconstruct the centralizing function of the recording studio glass, and the modularity of the creative practices of *Live* users extends itself to the environment of software programs made up of the computer hardware.

The *Live* environment's modularity is heightened by the development of standards for software interoperability, affordances that have facilitated greater interconnections between disparate media production software (figure 2.18). In addition to standard plug-in capabilities that allow for the incorporation of third-party instruments and effects into the *Live* workflow, the ReWire software protocol and Ableton's proprietary "Link" functionality

Figure 2.18 *Live* 9 promotional material, "Whatever You Want It to Be" (https://www.ableton.com/en/live/).

Figure 2.19 *Live* audio being sent to Avid's *Pro Tools* via ReWire (screenshot taken by author).

allow remote control and data transfer between *Live* and related software for digital audio editing (figure 2.19). For example, the audio output of a synthesizer patch in *Max* software can be "rewired" into the mixer of *Live*, allowing the musician to control effects and sound parameters of the *Max* synthesizer from within *Live*. Or a music-making app can be synchronized to *Live* on one's computer in order to align tempos and allow for remote interoperability. ▶ Since its emergence in 1998, ReWire has become an industry standard for music production software, allowing the simultaneous transfer of up to 256 audio tracks and 4,080 channels of MIDI data. In this way, software interoperability embodies what Manovich calls the "logic of 'permanent extendability'" in media production from the early twenty-first century.[59]

The convergence of instrumental and computational design models *within* the *Live* GUI and the proliferation of creative musical affordances *across* a larger network of software applications highlight an alternative understanding of the historical and technical relationships between media. If we approach software as an environment of affordances, we can think through how software encapsulates not only temporal relationships with previous media—the *instrumental* idea that technological change occurs through either innovation or remediation—but also an *accumulation* of technologies and creative practices from the past.[60] The quest for a platform that is not just one instrumental tool but can simulate many tools has been at the heart of computer science since its early days. In computer scientist Alan Kay's original vision of the computer, he characterized software as a "metamedium" that can "dynamically simulate the details of any other medium, including media that cannot exist physically. It is not a tool, although it can act like many tools." As such, the computer "has degrees of freedom for representation and expression never before encountered and as yet barely investigated."[61] The notion of the computer as a metamedium is both a challenge to existing practices of HCI and a metaphor for how the creative practices of one cultural arena can move across other platforms. In this mode of thinking, software environments are designed to facilitate an *accumulation* of tools and techniques rather than to serve only as a *remediation* of existing practices.

Thinking about the *Live* environment as a metamedium offers one explanation for the constant migration of compositional techniques and interface designs across media production software. For example, Manovich points out that the "layers" function, typical of photo editing software such as *Photoshop*, has become a standard technique across software, including vector image editors, motion graphics software, video editors, and sound editors. Despite the differences between media platforms, "a final composition is a result of 'adding up' data" for all these editing software.[62] Similarly, plug-ins and a modular GUI for sound manipulation, the core features of *Live*, serve to accumulate compositional techniques from new and existing instruments, as well as from tools and techniques that have migrated from other platforms. The additive nature of software development thus challenges electronic music producers to learn constantly new techniques and navigate new tools, encouraging them to focus on the *event* of creative production rather than the *objects* produced, thus destabilizing fixed understandings of musical instruments, musical "works," and software itself.

As I've suggested, the convergence of instrumental and computational design models exemplifies a shift from objects to processes in approaches to music making. By designing possibilities for extended functionality into the core mechanics of *Live*, the creators of the program posit new conceptions of the recording studio as a networked and "plugged-in" information architecture composed of tools and techniques gradually accumulated from external devices and applications. Peter Lunenfeld describes this process of technical accumulation as "stickiness," a fundamental quality of creative production in digital environments containing affordances "that allow other meaningful objects or systems to latch on to it, expand it, or burrow deep within it."[63] Software interoperability shows the extent to which a logic of stickiness is embedded within the design and development of software. In constantly seeking out perforations and "leakage effects" among the various spaces of the DAW, the interoperable nature of the software environment deconstructs the closed, insular walls of the physical recording studio.

The non-skeuomorphic GUI, modular organization of elements, and computational approach to design reflected in *Live* is meant to encourage more inclusive and accessible approaches to musical composition and production, thus carrying with it the potential to undo the monopolies of competence that have shaped the social and cultural politics of audio production for decades. Especially following the Me Too movement, this design philosophy of openness and modularity leaked outward from the technical details of the software to the marketing campaigns and social communities surrounding the software, resulting in an industry-wide push for accessibility and inclusivity in music production and audio engineering.[64] These efforts have been fostered by an entire economy of pedagogical materials designed for the purpose of instructing musicians on how to navigate new digital music production technologies, regardless of skill or background. Two parallel developments are especially significant: first, software marketing schemes that promote democratization and accessibility in the music products industry, and second, the emergence of digital music production schools and other educational content, particularly those catering to young women.

Since its first commercial release in 2001, *Live* has come to exemplify a DIY aesthetic of technological accessibility that characterizes the design, distribution, and use of the software. It's been at the forefront of user communities surrounding tools, techniques, and trends in digital audio production, often providing the headlines in magazines such as *Computer Music*, *Future Music*, and *Music Tech*, as well as online blogs such as Create Digital Music. These outlets simultaneously market new software and hardware to consumers, provide examples of technological use through artist interviews, and offer practical insights into how to use the tools through tutorials (figure 2.20). As such, they encourage broad accessibility and use of music software across amateur and professional demographics. The concomitant emergence of new techniques for sound manipulation and new forms of electronic music pedagogy further highlights the shift in conceptions of software from an archival medium onto which users inscribe a fixed *text* to a performance medium with which a new type of musician collaborates in the creative *process*.

In line with a marketing model based on the democratization of new technologies for music production, *Live* has ushered in a network of pedagogical practices to meet the needs of aspiring digital music composers.[65] In addition to the inclusion of *Live* in countless music technology courses at the university level, dedicated hybrid production schools have emerged, providing online and face-to-face instruction in a variety of digital audio topics. For example, Dubspot hosts courses on DJing, turntablism, sound design, mixing and mastering, and music theory and a special certificate program in *Live*. Instructors include professional DJs, producers, and sound engineers with specializations in specific software, as well as *Live* "Certified Trainers," educators who have successfully completed Ableton's in-house certification

Figure 2.20 *Live* 9 promotional material, "A Community"
Source: Ableton, "Live," accessed April 22, 2015, https://www.ableton.com/en/live/.

program. Together, the discourse and pedagogy surrounding digital audio production in the 2000s have been defined by an increased desire to market the software to as broad an audience as possible and to stabilize its use among instructors and artists across artistic disciplines.

Have these attempts at appealing to a broader user base actually paid off? In the years following the Me Too campaign, the music tech industry has been optimistic about the democratizing potential of certain techniques and technologies of music production.[66] Emerging both as pedagogical outgrowths of schools like Dubspot and responses to the lack of diversity within those spaces, international nonprofits in the 2010s attempted to diversify the music technology community by providing educational outlets and safe learning environments for young women and people of color. Beats By Girlz (BBG), for example, is "a non-traditional, creative and educational music technology curriculum, collective, and community template designed to empower females to engage with music technology," providing instruction to young women and nonbinary students around the world in music production, composition, and audio engineering skills.[67] In addition to hands-on instruction, BBG hosts collaborative workshops and networking events with major music tech companies. In her pioneering research on women in the music industry, BBG founder Erin Barra highlights the importance of networking opportunities for young women trying to break into the industry, and the social advocacy of BBG has resulted in countless networking events and training sessions for BIPOC and nonbinary Ableton users. As digital media artist Freida Abtan writes, "when people ask me how to get more women involved in electronic music culture, I have two answers: share your skills with them; but also: share your friends with them."[68]

The combined social and technical outlet provided by organizations like BBG is echoed in similar inclusion initiatives such as SoundGirls, Women in Music, She Rock She Rock, She Is the Music, and Women's Audio Mission, just to name a few. Laura Escudé, groundbreaking controllerist and the first Ableton *Live* Certified Trainer, provides another model for diversifying the field of music tech with the "Womxn Scholarship" offered by her group Electronic Creatives—a company that focuses on live tour playback engineering, advanced MIDI and instrument rigs, and content creation for live performances. These organizations have done extraordinary work bringing to light gender imbalances in the music industry, as well as pushing music technology companies to encourage broader representation in their user base. It's no coincidence that—in the wake of increasing connections between Ableton and groups like BBG and Electronic Creatives—Ableton responded to the demands of Me Too by increasingly featuring the work of women and POC artists on their blog, in their video tutorials, and in the programming for their annual Loop festival and music conference. It's yet to be seen how developments in the design of the software itself will reflect the increasingly "woke" culture that defined the 2010s. For example, how might the company rethink its expensive price point or the need for increased accessibility in design or diversity, equity, and inclusion (DEI) initiatives in the company's hiring structure?

As much as these initiatives reflect real progress and change for the music tech industry, some feminist thinkers and music technologists are critical of simply increasing the numbers of BIPOC and nonbinary users as a representational strategy. Feminist theorist Sarah Benet-Weiser and philosopher Robin James have both made the claim that increasing numbers of underrepresented people in an industry doesn't actually address structural problems in that industry.[69] As Born and Devine have shown in their research on music technology and education, it's possible for minorities to go into an industry and carry the same assumptions that they're supposedly trying to fight, echoing Ruha Benjamin's insights on the pervasiveness of white supremacy in software development more generally: "We could expect a black programmer, immersed as she is in the same systems of racial meaning and economic expediency as the rest of her co-workers, to code software in a way that perpetuates racist stereotypes."[70] Regardless of representation in the workplace, Benjamin also notes how it's naive to assume a given industry is so important that increasing numbers is all that's needed for change to happen. Going one step further, mixing engineer Leslie Gaston-Bird claims that the numbers argument can be misleading, since it buys into the long-held problematic assumption that "there are no women" in the industry, which isn't a productive way of thinking.[71] Speaking in the context of the television industry, producer Shonda Rhimes talks about how she's not trying to *diversify* television but to *normalize* it: "women, people of color, LGBTQ people equal way more than 50% of the population. Which means it ain't out of the ordinary. I am making the world of television look normal."[72] In light of these critiques, some have argued that increasing social networks for women remains the most effective strategy for DEI initiatives.[73]

Ultimately, the design elements in *Live* that I've addressed throughout this chapter reflect some core components of what Bardzell defines as "feminist HCI." Ableton pursues collaboration and equitable participation in user research, publicly demonstrates advocacy for communities of young women and people of color, and values self-disclosure through the use of computational design elements that "call users' awareness to what the software is trying to make of them."[74] We might also recognize, though, that while the computational design elements in software such as *Live* might offer a step in the right direction in terms of alleviating accessibility issues from a race and gender perspective, the skeuomorphic/flat design debate remains a Western, Eurocentric way of thinking about the "educational" potential of design. As Benjamin makes clear, "Design is a colonizing project, to the extent that it is used to describe anything and everything," and the so-called creative industries participate in the neocolonialist project as much as the technologies of cyberwar, racist AI assistants, and algorithmic surveillance.[75] Echoing the monopolies of competence in DAW design, she writes, "What, I wonder, are the theoretical and practical effects of using design-speak to describe all our hopes, dreams, qualms, criticisms, and visions for change? . . . Whether or not design-speak sets out to colonize human activity, it is enacting a monopoly over creative thought and praxis."[76] In the context of music software, specifically, DJ and music writer Jace Clayton highlights the fact that "virtually all music software is made in the United States or Europe. The programs all tend to do the same thing, in varying amounts, and that thing defaults to a narrow concept of what music can or should be." For Clayton, this point emphasizes his broader claim that "software tools are never neutral" and that "they reinforce their builders' blind spots and biases and, once widely distributed, play an active role in maintaining those assumptions."[77] As much as music tech initiatives like BBG hope for democratization, the reliance on *Live* as the core creative tool always presumes Western biases of musical knowledge. How might we get beyond this?

Design researcher Mark Blythe notes that a major priority of critical design involves "subverting and questioning norms through detournement, defamiliarization, and estrangement."[78] In designing his *Sufi Plug Ins*—"a suite of seven music-making devices based on non-Western conceptions of sound"—Clayton consciously fought against the values, assumptions, and roadblocks embedded in software such as *Live* in an attempt to craft a tool for "musicians from traditions not represented in software" (figure 2.21).[79] Four of the plug-ins reflect virtual synthesizers tuned to North African quarter-tone scales, while others include a clapping drum machine, an audio drone generator, and a device called "Devotion" which "lowers your computer's volume five times a day out of respect for the Muslim call to prayer."[80] In

Figure 2.21 Jace Clayton's "PALMAS" plug-in from the *Sufi Plug Ins*
bundle (2012)
Source: https://beyond-digital.org/sufiplugins/).

contrast to the user-friendly instrumental and computational design
approaches in *Pro Tools* and *Live*, *Sufi Plug Ins* are designed to frustrate typical
Western users, encouraging them "to explore the software's sounds guided by
their ears, with less focus on the numbers or language (music software loves
numbers)."[81] Regardless of what can seem like minute nuances in the design
of commercial DAWs such as *Pro Tools* and *Live*, both software programs reflect Eurocentric views of what's possible in digital music making. *Sufi Plug
Ins* provides a great example of what design researcher Milena Radzikowska
calls a "speculative feminist approach to design," in that it challenges existing
belief systems, seeks out what has been made invisible or underrepresented,
privileges transparency and accountability, and welcomes frustration and
critique in the UX.[82] As such, it presents a radical attempt at escaping the
Silicon Valley mold of innovation, progress, and the constant accumulation
of new features under the guise of user-friendly design.

In analyzing the isomorphic relationship between the technical design of
Live and the sociocultural network through which it proliferates, this chapter
has provided an account of how media production software in the 2000s
moves beyond and through the metaphorical perforations of the computer
screen. Perhaps more than any other software, *Live* is the tool of choice on the
stages of EDM festivals and multimedia art installations, as well as the *studios*
of film composers, video game sound designers, and "bedroom" electronic
musicians and DJs. A growing network of music pedagogues has contributed
to the widespread acceptance of the software by serving as mediator between
software designers and emerging practices of electronic music, helping
amateur and professional musicians to negotiate constantly changing design trends. The literal mobility and perceived accessibility of the software
throughout social and cultural spaces have contributed to the formation of

an aesthetic of technological accumulation and maximalist consumption that has defined media production since the new millennium.

However, the maximalist aesthetics embedded within these plug-in cultures aren't the only modes of computational practice available to electronic musicians, software designers, and instrument makers. As much as the maximal interface of the DAW evokes the promise of unlimited creative options through the additive plug-in mechanics embedded within the seemingly transparent software, others perceive this influx of prescribed functionality to be creatively limiting. Describing what he calls the "deadly embrace between software and users," Miller Puckette claims that "no matter how general and powerful we believe today's software to be, it is in fact steeped in tacit assumptions about music making that restrict the field of musical possibility."[83] From this perspective, it makes perfect sense that the festival culture of EDM—with all of its sonic grandiosity, digital bombast, and excessive "drops" that overwhelm the sensorium—has become the genre par excellence for users of software such as *Live* and *FL Studio*. Whereas Dennis DeSantis and other electronic music pedagogues are concerned with introducing new compositional mindsets to electronic musicians in an attempt to help them navigate the inherent complexity of the DAW, Puckette argues that "an even more powerful strategy for managing complexity is simply to avoid it altogether."[84] Chapter 3 tests this minimalist proposition by reversing the questions posed by the design of *FL Studio*, *Pro Tools*, and *Live*: rather than being presented with an influx of creative options in a maximal interface, how have electronic musicians approached software that seems to provide no options at all? What's the nature of music production, instrumentality, and musicianship when all you're presented with is a blank screen?

3

Terminal Aesthetics

Software developer, music educator, and mathematician Miller Puckette is first and foremost a pragmatic guy. When it comes to technology, he values limited tools that do seemingly unlimited things, embracing a "less is more" mentality in both his professional work and everyday life (figure 3.1). In opposition to the "upgrade culture" of Apple and the rest of Silicon Valley, Puckette finds a Zen-like utility in technologies that do the job right without getting in the way. Guided by what he calls the "universal principles of computer science"—portability, abstraction, reusability, and interoperability of code—Puckette's research at IRCAM in the late 1980s led to the development of *Max*, a pioneering software for process-oriented electroacoustic music performance, as well as *Pure Data* (*Pd*), its open source equivalent. Puckette is distinctly critical of how emerging tools for music production guide the user down predefined paths through flashy interfaces and novel mechanics, and software in the "*Max* paradigm," including *Max* and *Pd*, instead presents the user with the digital equivalent of a blank canvas: an empty screen.[1]

While the idea of building an entire working system from the ground up is a familiar challenge to computer programmers, it's mostly new to digital musicians used to working with the familiar studio interface metaphors of knobs, sliders, and linear audio editing discussed in chapter 2. How can we talk about the affordances and constraints of musical interfaces in the context of software that leaves the primary design responsibilities to a user with a blank screen? In other words, what's the nature of musical instrumentality in the age of computational thinking? Further, what's the relationship between the terminal aesthetics of programming sound and the cultural politics of race and gender in the history of computer programming?

Not simply a tool for recording music, *Max* is a visual programming "environment" for music production that's become increasingly popular among digital artists since its first commercial release in 1990.[2] While many commercial DAWs present the user with a linear, timeline-based interface best suited for music *recording*, *Max* specializes in process-oriented digital art, presenting the user with a blank screen onto which various interactive

Push. Mike D'Errico, Oxford University Press. © Oxford University Press 2022.
DOI: 10.1093/oso/9780190943301.003.0004

Figure 3.1 Miller Puckette installs *Pure Data* on a Raspberry Pi
Source: "Miller Puckette: Pure Data on the Raspberry Pi," Alexander Matthews, October 26, 2012, accessed August 27, 2020, https://vimeo.com/52265243.

"objects"—building blocks used to create programs in *Max*—are added. Many have lauded the open, flexible, and adaptive nature of the software, often comparing the creative experience of *Max* to the seemingly all-access control one has while programming computers. Yet, while artists have been quick to praise the effects of this software on their own work, the procedural and computational aesthetics inherent in the program are rarely contextualized outside of these communities of practice.

Through an analysis of the technical design of *Max*, this chapter offers insights into how ideas about musical instrumentality might be expanded as a result of the increasing integration of music and computation. As music composition increasingly converges with software, instrument designers develop interfaces that borrow from a range of design disciplines, constantly introducing new sets of creative affordances for musicians and composers. I align these new forms of digital literacy with the concept of computational thinking, a theory that emphasizes not only how individuals learn how to use computers but also how computers themselves shape individuals' understandings of culture and creativity. In applying these principles to musical practices, I introduce the practice of *procedural listening* to describe how electronic musicians come to learn musical concepts and techniques while engaging in nonlinear computational processes. Procedural listening represents a new form of instrumentality in the digital age: a primary skill developed by audio producers that allows both musicians and audiences to focus on the process-oriented mechanics of media *forms* rather than audio *content*. As musicians embrace procedural listening in order to understand the inner workings of software, their understanding of musical instrumentality shifts from a *tool*-based model, in which the instrument is a concrete means to reach a specific musical goal, to a *system*-based model, in which the instrument is part of an integrated technological network used to explore previously media-specific creative possibilities.

On Monday, December 8, 2014, President Barack Obama made history in becoming the first president of the United States to write a line of computer code. During a promotional event for Computer Science Education Week, the president sat down with a group of middle-school students in New Jersey to speak about the importance of computer science in public education, following a major announcement that the White House would be donating more than $20 million in contributions to Code.org toward its efforts to train ten thousand teachers in computer science education by the beginning of fall 2016 (figure 3.2). Obama's coding session was just one of many media events to promote the "Hour of Code" campaign, an international nonprofit developed to instill core values of computer science, including technological

Figure 3.2 Coding session with President Barack Obama (2014)

Source: Klint Finley, "Obama Becomes First President to Write a Computer Program," *Wired*, December 8, 2014, accessed August 27, 2020, https://www.wired.com/2014/12/obama-becomes-first-president-write-computer-program/.

accessibility and digital literacy.[3] Since the program's launch in December 2013, various celebrities—from Silicon Valley giants Mark Zuckerberg and Bill Gates to actors Ashton Kutcher and Angela Bassett—have expressed support for the campaign, promoting the value of computer coding to grade-school students, soldiers, and doctors alike.

The "Hour of Code" is an example of a twenty-first-century Big Tech initiative that perpetuates a specific brand of digital literacy.[4] The bolstering of STEM and computer science education in public schools runs parallel with the proliferation of three cultural communities that foster similar values: the free software movement, maker culture, and live coding. These scenes exemplify how the push for digital literacy has ushered in more relational forms of HCI in the practices of computer engineers, media artists, and musicians. If the previous archetype for computer programming was the lone hacker, in full control of the software in front of him, these new models conceptualize computing as a two-way, relational process of negotiation between the knowledge of the individual and the affordances of the software. This ongoing process of negotiation—the push and pull between software and its users—is fundamental to the practice of procedural listening.

Since the early 1980s, the "free software" movement has advocated for the uninhibited use, distribution, and modification of software. Computer programmer Richard Stallman's "GNU Project"—an attempt to create an operating system composed entirely of free software—established the values of this movement in direct opposition to the proprietary ownership model enacted by major technology corporations such as Apple and Microsoft. Table 3.1 outlines commonly perceived differences between the two models of software distribution.

Table 3.1 Comparison of "free" and "proprietary" software cultures.

Free Software	Proprietary Software
Free to use	Pay to use
Online distribution	Commercial distribution
Source code available for use	Source code unavailable for use
Emphasizes "hackability"	Emphasizes "usability"
Requires more time to learn	Requires less time to learn
Smaller development team	Larger development team
Anarchist-libertarian ethic of use	Democratic ethic of use
Values "freedom"	Values "innovation"

The differences in software *design* between the two models come down to the question of "hackability" versus "usability." In the case of free software, does the software offer greater possibilities for modification at the expense of being user-friendly? Or in the case of proprietary software, is the software accessible to a greater range of users at the expense of being open to modifications? At the root of these questions is digital literacy.[5] In order to embrace the full capabilities of free software, the user must overcome a learning curve that often involves the development of computer coding skills. The average computer user may not be willing to dedicate the time and energy to this task. In contrast, advocates of free software see digital literacy as not only a technical challenge but a political and ethical responsibility for reclaiming control of digital culture from the large corporations of Silicon Valley.

With the introduction of cheap microcomputers such as the Raspberry Pi in the early 2010s, the free software movement converged with a range of DIY creatives to form what's become known as "maker culture." This informal network of computer hackers, robotics enthusiasts, 3D printers, and artisans working in the traditional arts and crafts has been guided by the principles of learning through doing, hacktivism, and shared knowledge formation (figure 3.3).[6] Significantly, maker culture inspired an even greater push for computer science education in America, as microcomputers continue to be used as tools in teaching children how to code. Similar to the free software movement, makers see themselves in opposition to corporate ideologies of technological "innovation" and "progress," instead valuing knowledge formation from the ground up through amateur production practices.

Figure 3.3 Paris Maker Faire, 2015

Source: "35,000 visitors at the Paris Maker Faire, a record," *Makery*, May 4, 2015, accessed August 27, 2020, http://www.makery.info/en/2015/05/04/35000-visiteurs-a-la-maker-faire-paris-un-record/.

Finally, digital literacy has been given a performative dimension by computer programmers who engage in "live coding," projecting their computer screens to an audience while they code software in real time.[7] Live coding is practiced in a range of art forms, including dance, poetry, and music, typically highlighting the technical virtuosity and improvisational skills of the programmer (figure 3.4). In addition to its entertaining and community-building functions, live coding can be educational. Online streaming hubs, such as livecoding.tv, often host instructional tutorials to help viewers learn computer programming techniques.[8] The combined performative and pedagogical functions of live coding foreground a community advocating for digital literacy by showing audiences the process of coding, rather than just sitting behind a screen and "pressing play."[9]

The free software movement, maker culture, and live coding reflect the increased presence and influence of computer science principles on twenty-first-century digital culture. By digging behind the technical practices at work in computational thinking communities, it's possible to see how their countercultural politics of technological alterity to Silicon Valley have become fixed within the cultural practices of DIY computing.[10] However,

Figure 3.4 Live coding session in "La Fabrique" (Nantes, France, 2012)
Source: "Live coding session in La Fabrique (Nantes, France)," n.d., accessed April 29, 2016, http://diccan.com/Eicl.htm.

despite clear attempts by these communities to externalize the mechanics of software to their audiences, tracing cultural aesthetics as they manifest in the design of software itself is not so obvious. This chapter outlines how software in the *Max* paradigm epitomizes the principles of digital literacy, principles that drive the creative practices of digital cultures. As both musical instruments and forms of musical notation, these tools introduce unique forms of HCI, encouraging composers to become designers, designers to become listeners, and listeners to become performers.

The history of the *Max* paradigm epitomizes the dichotomies between free and proprietary software previously discussed. Puckette first conceptualized ideas for a real-time computer control solution for music in the 1980s at IRCAM. Throughout its history, IRCAM has perpetuated modernist ideologies that value newness and innovation above all else, including "a hostility and contempt toward all commercial developments and especially 'low-tech' or small consumer technologies," according to Georgina Born.[11] Software was a specific target, "denigrated for having no such physical embodiment, no object form, for being insubstantial and ephemeral."[12] As a result, most of the software developed at IRCAM after the advent of digital computers involved highly opaque programming languages that required specialized knowledge in computing, acoustics, music composition, and electronic music.[13] Even in the realm of hardware, Pierre Boulez—the director of IRCAM—reportedly told the pedagogy director that the user-friendly Apple Macintosh computers would come into the institute "over my dead body."[14]

Max represented a significant departure from the modernist ideology of IRCAM. First, it's a software with a GUI known as a "patcher," rather than a text-based programming language (figure 3.5).[15] The patcher design afforded more opportunities for real-time music making, which was the primary goal of the software. Second, *Max* was designed as a performance tool, rather than a music "compiler" whose purpose was to produce notated scores or audio recordings. Despite the advent of commercial digital sampling hardware such as the Fairlight computer in the late 1970s, which allowed for a certain level of "real-time" audio editing, the anti-commercial stance of Boulez caused IRCAM to rely instead on outdated machines. As in the early days of computer music, to hear brief snippets of their music, composers had to wait hours—sometimes days—for the sounds to be compiled by the machine. In addition to the software's emphasis on performance and its use of a GUI, the *Max* design also departed from IRCAM ideology by integrating the

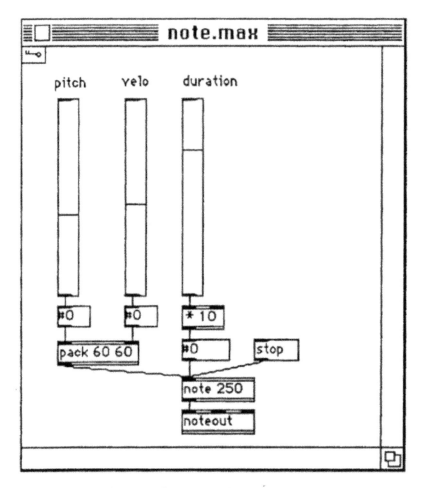

Figure 3.5 Original *Max* patch prototype (1988)

Source: Miller Puckette, "The Patcher," in *Proceedings of the International Computer Music Association* [San Francisco: ICMA, November 8, 1988], http://msp.ucsd.edu/Publications/icmc88.pdf.

"commercial" Apple Macintosh into the performance setup, eventually enabling the first version of *Max* to be used onstage.[16]

The discrepancy between user-friendly and programmable software design played out not only in the development of *Max* at IRCAM but also in its commercial distribution beyond the institute. In 1989, IRCAM licensed the software to Opcode Systems, which sold the first commercial version in 1990. In 1996, Puckette developed *Pure Data* (*Pd*) as an open source alternative to *Max*, in an effort to make several improvements to the program. Since then, *Pd* has become central to the free software movement. Meanwhile, *Max* has followed a more proprietary route, aligning itself with popular commercial software such as *Live*, and incorporating a more plug-in-based interface design (Ableton acquired Cycling '74, the developers of *Max*, in 2017).[17] Despite these differences, the *Max* paradigm continues to offer modular platforms for creating process-oriented musical experiences to digital artists across media, sporting a "visual programming" interface design that encourages users to develop digital literacies more akin to computer programming than to musical composition.

The capabilities for interfacing among hardware devices and software applications have inspired both the creative practices of *Max* users and their conceptions of music composition in digital environments. Among electroacoustic composers and others working in the Western art music tradition, *Max* has been used to supplement acoustic performance with real-time and improvisational electronics. In an interview with Cycling '74, composer William Kleinsasser talks about how the software introduces new concepts and methods into the creative process. Describing the rise of what he calls the "integrative composer," Kleinsasser claims that "many composers now speak of a blurring difference between composition and programming."[18] The rise of the composer-programmer is a reflection of how the digital literacy afforded by *Max* has introduced modular, systems-based approaches to the compositional processes of Western art music.

Similarly, among electronic producers working in popular music, the modular approach to composition in *Max* is embraced as a way of pushing their music into more experimental territory. In his work with Radiohead, Jonny Greenwood has used *Max* as a respite from "all those programs [that] seemed desperate for you to write in 4/4 at 120 bpm and loop the first 4 bars."[19] Los Angeles beatmaker Daedelus uses the software in conjunction with hardware interfaces to stretch, stutter, and juxtapose fragments of existing tracks, *Max* being used as a hypersampler of sorts (figure 3.6).[20]

While the software has been diversely employed for the creation of music, other users have integrated *Max* into multimedia applications. In the Skube media player, Andrew Spitz created a Spotify radio by combining an Arduino board, *Max* software, and the Last.fm API.[21] Skubes exist as physical, cube-shaped speakers that react to being flipped, tapped, and connected to each other by either shuffling music from a playlist or selecting music based on songs previously accessed by the listener (figure 3.7). In this case, *Max* is used

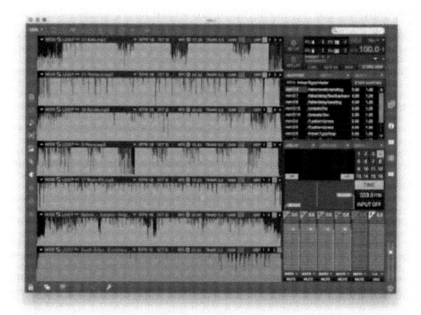

Figure 3.6 MLRv 2.2 (2011), a *Max* patch by Galapagoose, allows users to simultaneously edit and play back multiple audio loops (screenshot taken by author).

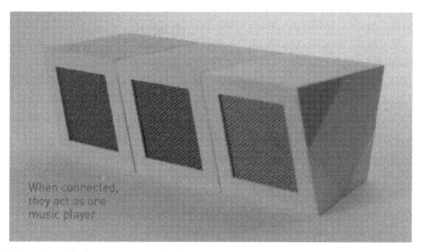

Figure 3.7 Skube promotional video

Source: Andrew Spitz, "Skube: A Last.fm & Spotify Radio," September 12, 2012, accessed August 27, 2020, https://vimeo.com/50435491.

both to interface with the physical cube through the Arduino and to control audio playback in Spotify through its integration with a custom-coded Applescript.

Using music software in the ways just listed requires computational forms of digital literacy that privilege system-oriented thinking, technological interoperability, and abstract problem-solving. In short, the *Max* paradigm encourages music *composers*—in the broad sense of the word—to think like computer *programmers*.

If digital literacy focuses on the *processes* through which humans learn how to use computers, computational thinking constitutes the *attitude* that individuals hold toward creative production with digital tools. Computer scientist Jeanette Wing defines computational thinking as a "fundamental skill for everyone, not just for computer scientists," which involves "solving problems, designing systems, and understanding human behavior by drawing on the concepts fundamental to computer science."[22] For Wing, computational thinking is more than just the ability to program a computer; it requires the skills to work abstractly with entire systems, rather than just their individual elements; to reframe and reformulate difficult problems into simpler language through simulation; to think through both theoretical and practical solutions to a given problem; and, most significantly, to discover solutions in the presence of uncertainty.

The skills reflected by computational thinking are useful in understanding not only digital interfaces but also the mechanics of musical instruments and compositional systems. Music educator Gena Greher and computer scientist Jesse Heines introduce the idea of "computational thinking in sound" as a teaching tool for music and computer science students, both of which are "hampered by habit, which limits their abilities to imagine alternative possibilities."[23] For example, the ability to view musical structures as a hierarchical set of data allows students to understand melody, lyrics, and song form as chunks of connected data from which they might extract meaningful patterns. Greher and Heines offer activities to help students "crack the code" of musical notation (Western scores, MIDI piano rolls, audio waveforms, etc.), encouraging the development of abstract notation schemes that rely less on systems of representational metaphors that have become reified throughout the history of Western music and more on the counterintuitive mechanics and awkward physical gestures required of students when they first encounter a new instrument or interface.

The efficiency of well-constructed code, the ability of software to evolve in relation to its hardware environment, the conceptual abstraction of code into a series of programmatic steps, and the recursion and repetition of digital processes are just a few of the values that continue to surface in computer science theory and pedagogy.[24] These values constitute an aesthetic of minimalist elegance and interoperability in the construction and operation of computational systems. Software developers, game designers, and electronic musicians alike apply these values in their working methods, which then get inscribed into the design of software. Among the many principles of computer science, *proceduralism* has become an influential tool for thinking about the technical *processes* behind software's inner workings. Computer scientist Michael Mateas describes proceduralism as "the ability to read and write processes, to engage procedural representation and aesthetics," and to understand code as a written artifact with its own embedded aesthetics, language, and poetics.[25] Game designer and theorist Ian Bogost defines "procedural rhetoric" as a "process-intensive" engagement with computational systems, in which "expression is found primarily in the player's experience as it results from interaction with the game's mechanics and dynamics."[26] To think procedurally is to think *in* and *through* algorithms and mechanics, as tools for critically engaging computational procedures both as objects of analysis and as interactive experiences. Procedural literacy and procedural rhetoric provide examples of how the process-oriented nature of computing has fostered computational thinking across media cultures. In turn, computational thinking has directly shaped existing ideas about creativity, including temporality in formal structure, the nature of representation, and issues of authorship.

Concentrating on the underlying logic of software requires a perceptual shift away from the narrative capabilities of media toward the nonlinear relationships between discrete elements within larger systems. Bogost describes these aspects of computational systems as "unit operations . . . modes of meaning-making that privilege discrete, disconnected actions over deterministic, progressive systems."[27] Considering software at the unit operations level shows how computational systems are designed to be more than just metaphorical representations of existing phenomena. Examples include a computer programmer deciphering software by picking apart the individual lines of code or a video gamer navigating an abstract world by learning its unique rules and physics (see chapter 6). Systems design engineer and theorist Wendy Hui Kyong Chun claims that the interactive elements of computational systems are precisely rooted in this ability of software to break down mathematical operations into a series of simple arithmetical steps, as she claims, "The programmability and accuracy of digital computers stems from the discretization (or disciplining) of hardware."[28] The "discretization" of digital audio production into modular "objects" has allowed for new forms of computational thinking to influence electronic music composition.

In addition to altering musicians' and listeners' sense of temporality in the formal structures of creative work, proceduralism evades the representational power of digital tools, instead focusing on the power of abstraction afforded by the computer. Rather than using text description or image depiction to explain relationships between objects and events, proceduralist media make claims about how things work by modeling them in the process-oriented environment of the computer.[29] Whereas print writing and other textual media excel at *describing* phenomena, and visual media *depict* that same phenomena, software *synthesizes* cultural information of various sorts, including the compositional processes of music.

Finally, as the experiential focus of digital media shifts from objects to environments, computational thinking reconfigures existing understandings of authorship. With "analog" media, it's conventional wisdom to think of the author as the person who writes poetry, composes music, and so on. Instead, software encourages its users to author *new processes* as a fundamental aspect of media creation.[30] This is a significant point for music production, as it allows the producer to move beyond the idea of music software being strictly designed for the recording of audio content to a fluid conception of software as a real-time instrument for the performance of music that's not yet complete.

A core element of computational thinking, proceduralism offers insights into how users directly interact with the mechanics of software. At the same time, the feedback loop between a user and a given technology also involves moments of role reversal, in which the user takes a step back and listens to the machine. What's the nature of those experiences in which a computational process is set in motion and the "programmer" becomes a seemingly uninvolved observer? What happens to the subjectivity and agency of computer musicians after they press play? The composition of digital music involves active moments of sound recording and editing and observational moments of listening and analysis.[31] I describe this element of the digital music production process as *procedural listening*, in which the creative focus of the producer shifts from the audio *content* being created to the *formal* mechanics of the computational system. Thinking about music production as a type of procedural listening allows us to see how digital and musical literacies are not simply about learning how to create music with computers but also about learning how to interact physically and cognitively with new instruments, technologies, and cultural practices. Music composition and HCI are typically thought of as top-down relationships of a human "user" providing one-way inputs or "commands" to a computer. Procedural listening complicates this dynamic by focusing on how the creative input of "users" is guided conversely by the "rules" of the computational system, thus fostering more relational forms of HCI.

Procedural listening is not limited to software-based music. It's also present in many twentieth-century art forms, including the process-oriented aesthetics of the post-1960s experimental music traditions. In response to the creative affordances for music performance offered by technologies such as electromagnetic tape, minimalist composer Steve Reich developed the idea of "music as a gradual process," referring not to "the process of composition itself, but rather pieces of music that are, literally, processes."[32] Composer Brian Eno expanded on Reich's ideas in the context of the recording studio, in a creative process he calls "generative music": "the idea that it's possible to think of a system or a set of rules which once set in motion will create music for you."[33] While Reich was concerned with how technology influenced the performance practices of human musicians, Eno focused on the influence of machine processes on other machines (figure 3.8). These examples, among others, serve not only to detail a rich lineage of music as algorithmic process

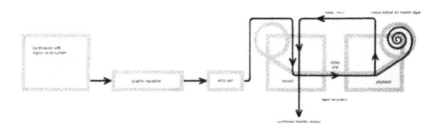

Figure 3.8 Operational diagram for Brian Eno's *Discreet Music* (1975). To the left, a delay loop is created as tape recordings play back at different speeds.

Source: Michael Peters, "The Birth of Loop," *Prepared Guitar*, April 25, 2015, accessed August 27, 2020, http://preparedguitar.blogspot.com/2015/04/the-birth-of-loop-by-michael-peters.html.

but also to highlight how cultural responses to technological changes are historically specific. In this way, it's possible to understand forms of musical instrumentality as co-constructors of social and cultural eras.

As the examples of process-oriented music attest, it's the element of perceived randomness in the perception of procedurally generated audio events that allows musicians to apply procedural listening in their creative approach. Electroacoustic composer Kim Cascone talks about how satisfying it is to be able to "get dealt all these random events and try to make sense of it on the fly . . . it kind of develops a *certain way of thinking* about the material."[34] Computer artists of the 1960s used random number generators both to break the predictability that came from the human influence on computer programming and to develop programs that could replicate the work of existing artists by simulating their "rules" and patterns. Dealing with randomness and complexity is also a concern for computer scientists, whether in attempting to predict the uncertain results of algorithmic procedures or network shifts or in theorizing the complexity of computational systems themselves.

The "certain way of thinking" about generative music described by Cascone involves a shift in creative roles encouraged by procedural listening, as the musician learns to think like a designer rather than a composer. Most noticeably, it involves a move away from a "composerly" perspective, as musicians learn to forgo compositional values such as intentionality and authorial control over the material.[35] Singer, songwriter, and audio engineer Ducky views herself as a *designer* of the musical process rather than a composer as such: "Something that I want to play with more is the randomization of things, something that I find really beautiful in sets. Maybe I can control a few things and set up whatever patches and plugins and stuff so that it's something that grows in a way where I'm not in control of everything."[36] Here music composition becomes a balancing act between production and consumption, performance and observation, compositional control and computational agency.

Redirecting listening practices from "musical" content such as pitch and harmony to the organizational principles of computer code requires computer musicians constantly to develop and redevelop digital literacies. Engaging conversations among interface designers can help us understand how the literacies required of procedural listening—the ability to understand the abstract nature of algorithms; competency in decoding unit operations and other discrete elements of computational systems; process-oriented thinking—are fostered on a cognitive and practical level.

Music software, like all consumer products, is the result of a long and complex *design* process. As such, every software program contains within it social, cultural, technological, economic, and aesthetic values that fade to the background of musicians' attention the more they use the software. To consider the values fixed within the software, we need to decode the design philosophies that lead to the widespread use of the program.

Popular trends in design research have been concerned with how best to achieve a "natural" interaction between humans and technologies, thus eliminating the need for users to develop new literacies for every technology with which they engage. Standard designs for mundane tools such as doorknobs, kitchen utensils, and vehicle dashboards are just a few examples of this push toward more "intuitive" forms of HCI. Design thinker Donald Norman proposes not just design principles for practitioners but an entire "psychology" of how individuals interact with everyday things.[37] For Norman, the goal is "natural design," in which users don't experience their interaction with technology as an obstacle, instead focusing on the task that needs to be completed at any given moment.

A quintessential element in natural design is maximizing the transparency between the intended use of a device and its user. Norman defines the mapping between intended actions and actual operations as interface "visibility," claiming that "just the right things have to be visible: to indicate what parts operate and how, [and] to indicate how the user is to interact with the device."[38] Consider a standard, sixteen-channel audio mixing board. The device is designed to expose as many gestures as possible that a user might make in manipulating sound, from turning knobs to fading sliders and plugging in cables. These interface elements are openly laid out on a flat surface, often color-coded to further highlight and distinguish actions, in an attempt to make the process of audio mixing as transparent as possible. The board is designed to make musicians feel they're not interacting with a mixing board at all but rather touching and manipulating sound itself.

By mapping technological operations in a transparent manner, visibility acts to interface designer and user. With a clear knowledge of the intended use of a given software, the user can develop a mental model that simulates the inner workings of the program.[39] The more the designer exposes the properties that determine how a program could possibly be used—what Norman calls "affordances" and "constraints"—the more fully developed the conceptual model will be.[40] The thought process behind the conceptual model is this: with a knowledge of how a device *should* work, users will ideally forget that they're confronted by a technology altogether, as the interaction becomes naturalized into their everyday practice.

In addition to conceptual models, natural design can be accomplished through the use of skeuomorphs. In software operating systems, for example, tabs, folders, desktops, and other design metaphors are employed as navigational guides that are presumed to be understood intuitively by the broadest range of users. In its "iOS Human Interface Guidelines," Apple suggests the use of interface metaphors as a strategy for increasing the feeling of transparency between users and mobile apps: "When virtual objects and actions in an app are metaphors for familiar experiences—whether these experiences are rooted in the real world or the digital world—users quickly grasp how to use the app. It's best when an app uses a metaphor to suggest a usage or experience without letting the metaphor enforce the limitations of the object or action on which it's based."[41] While Apple is quick to recognize the creative constraints and limitations inherent in the use of design metaphors, the marketing and use of "everyday" technologies such as Apple products rely heavily on the user perceiving an unmediated, or "invisible," relationship to the device.

Here's the major limitation of the natural design philosophy. In emphasizing user *experience* over the technical *materials* that make up the interface, natural design implicitly values the *mundane* forms of technological interaction that define the everyday actions so clearly delineated by Norman and others. In this way, natural design philosophy risks losing sight of a core aspect of HCI: the *creative* ways in which artists (as well as many "everyday" users) skillfully navigate and experiment with the material structures of their devices. While HCI experts and designers continue to believe that the ideal HCI should be invisible and get out of the way, early computing pioneers originally conceived of the GUI as a *medium* designed to facilitate learning, discovery, and creativity.[42] Norman provides useful ideas for designing objects with common usage conventions, but natural design principles don't easily map onto music production software designed to generate entirely new creative experiences. To understand how music software can facilitate new digital literacies through computational thinking, it's important to look beyond philosophies of *interface* design toward principles of musical *instrument* design.

In many ways, everyday UIs and common musical instruments share natural design principles. Just as the design of a doorknob rarely departs from a handheld object that rotates in a circular fashion, so, too, does the design of an electric guitar rarely depart from a long wooden fret board attached to a body on which the musician can rest a hand. Over decades of use, the cultural techniques and physical affordances of these tools have become transparent to their users, resulting in a direct mapping between the intended design and the actual use of each technology.

In contrast, electronic musical instrument designers have thought about how the rise of alternative interfaces complicates traditional notions of cognitive mappings between intended and actual musical operations.[43] Instrument designer Joseph Butch Rovan brings up the importance of *resistance*, rather than transparency, as an aspect of HCI in the context of music. Quoting Aden Evens, Rovan claims that "the resistance of the interface 'holds within it its own creative potential . . . the interface must push back, make itself felt, get in the way, provoke or problematize the experience of the user.' "[44] By considering the interface an obstacle—a problematization of UX rather than a crutch—Rovan provides a counterargument to the "natural design" principles previously laid out. Similarly, programmers Chris Nash and Alan Blackwell describe how reducing the "closeness of mapping" between interface metaphor and conceptual model introduces a shift from design principles based on *usability* to those that encourage *virtuosity* in music systems.[45] Exposing how interfaces "get in the way," or perhaps even "create the way," rather than gradually fading away and becoming "invisible" to the user, instrument designers provide productive ways of understanding design as a tool for introducing new digital literacies to musicians and composers.

Musicologists have taken this idea a step further, asking how epistemological definitions of "music" are bound up with the technical design of musical interfaces. In her article "Toward a Musicology of Interfaces," Emily Dolan defines "instrumentality" as the modes of mediation at work in the technologies that enable musical production. Historicizing the development of the keyboard interface in line with rational thought and enlightenment aesthetics, Dolan investigates how "the keyboard has represented a particular mode of instrumentality, namely one based on the idea of complete control."[46] If modernity presents a unique form of instrumentality shaped by a specific historical context, procedural listening recognizes the diminishing value of technological control as a defining feature of musical instrumentality in the age of computational thinking.

Understanding the concept of design "affordances" is crucial in explaining how interfaces might resist the control of users. For the most part, designers and thinkers have used the affordance concept as a springboard to detail the constraints of technological systems, focusing on how artists often use technology in ways unintended by the designer. But how can we talk about the affordances and constraints of musical interfaces in the context of software based on a "terminal" design, that is, software that leaves the primary design responsibilities to a user simply presented with a blank screen? What types of digital literacy arise when a user is encouraged to build procedural, dynamic systems with seemingly no feedback from the computer?

Current understandings of interface and instrument design are limited in that they focus on the control of the user over how the technological system could possibly shape the creative mindset of the user. On one hand, interface designers too often focus on how design should embrace affordances that have been cultivated by users in the course of a product's history. On the other hand, instrument designers can be rightly accused of the opposite: emphasizing how users productively *subvert the constraints* of a technological system. In either case, the creative focus remains on the desires and intentions of perceived users working to shape their experience. How might forms of digital music literacy change if we reverse this model, examining how users act not simply as agents of control over technological systems but also *in response to* the inscribed codes, algorithms, and computational procedures of those systems?

While software such as *Pro Tools* may be better suited for musicians looking to record through-composed tracks, often via the simulation of existing instruments and tools, software in the *Max* paradigm specializes in the creation of instruments and compositional *processes* themselves. Upon launching the program, the user is presented with a blank screen similar to the command line interface of the Mac operating system (figures 3.9, 3.10, 3.11).

As a self-proclaimed "visual programming language for media," this blank screen design aligns the *Max* paradigm with the "terminal" interfaces of many computer programming languages. The terminal design philosophy privileges bottom-up creative environments in which the programmer-artist can create comprehensive systems or tools by working through the interactions of the smallest possible units. Jonny Greenwood talks about how the "open" nature of the blank *Max* screen allows him to break free from the seemingly strict conventions of software like *Pro Tools* which makes him feel like he's "always being led down a certain route." When he started using *Max*, he claims "it was like coming off the rails. Before there was all this padding between the computer and me. Now there was a blank screen as a starting

Figure 3.9 *Pure Data* blank screen (screenshot taken by author).

point."[47] The DIY feeling of immediacy toward the *Max* interface is pervasive among users and reflects an increasing trend toward alternative digital literacies in the development of procedural listening.[48] ▶

In the case of *Max* and *Pure Data*, the terminal design is fostered through operational "commands" in the GUI known as "objects." These modular units can be added to the blank interface through an "object browser." Some of the more common objects include a "button," which simply sends a "bang" (a trigger to do something) to another object; "metro," which sends bangs at regular intervals, most often used to create a metronomic counter; and "toggle," a switch to send on/off messages to other objects. There are approximately seven hundred objects that come bundled with the commercial release of the software, each accomplishing a wide variety of tasks, from manipulating the GUI of the user's project to performing arithmetic and displaying visual media (figure 3.12).

While the objects alone are useless, connecting various objects together through virtual cables can create complex musical applications. As a basic example, figure 3.13 shows the *Max* "patch" for a random note generator.

In this patch, the square "toggle" (1) at the top triggers the following progression of actions: a "metro" (2) begins counting every second (1,000 milliseconds), which triggers the circular "bang" (3) at each interval. This "bang" tells the "random" message (4) to generate a MIDI note from 0 to

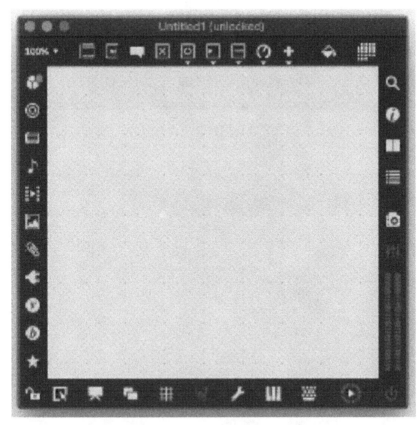

Figure 3.10 *Max* 8 blank screen (screenshot taken by author).

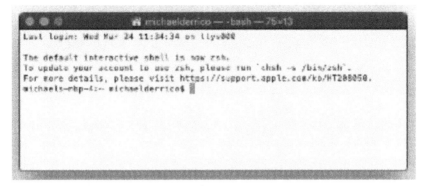

Figure 3.11 Mac OS X command line interface (screenshot taken by author).

Figure 3.12 *Max* 8 object browser (screenshot taken by author).

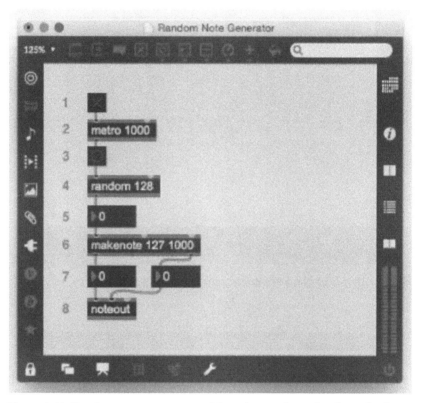

Figure 3.13 Random note generator in *Max* 7 (screenshot taken by author).

127, a value that gets displayed in the "integer" box (5) just below. The MIDI number is then sent to a "makenote" object (6), which adds a velocity of 127 and a duration of one second to the MIDI value, thus creating a full set of MIDI parameters. The note and velocity (7) are finally sent to a "noteout" object (8) that transmits the numbers to a designated MIDI device (often the computer's built-in synthesizer), resulting in a sounding note every second.

This sequence of actions continues as long as the "toggle" is on, designated by an X across the box. ⊙

There are three main elements of procedural listening at play while working with the *Max* paradigm: (1) "incomplete thinking" as a systems-based approach to composition; (2) the employment of rapid prototyping into one's creative workflow; and (3) the ability to integrate the mechanics and underlying structures of other media platforms into the composition and design of musical processes.[49] These skills highlight the increasing convergence of computation and composition, as well as providing analytical models for understanding process-oriented digital art. In his 2002 essay "How I Learned to Love a Program That Does Nothing," *Max* developer David Zicarelli defines "incomplete thinking" as a strategy for artists looking to begin working with the program. Rather than conceptualizing final goals through the realization of compositional intentions, *Max* users must "think about the middle instead of the end, in the same sense that the programmer of a word processor is more concerned with how documents are edited than the quality of the writing being composed."[50] The idea of incomplete thinking as a focus on the *how* rather than the *what* resonates with much of the previously detailed proceduralist theories and expanded notions of process-oriented instrument design.[51] Whether in the design, marketing, or use of *Max*, the idea of the computer software as a real-time performance environment in dialogue with its user is ubiquitous and has shaped both how new users learn the program and the increasing convergence between artists working across media.

The incomplete aspects of the *Max* paradigm are not simply a result of the tools afforded by the software but also a product of the relationship between the minimal, "terminal" design of the interfaces and the structure of computer programming languages. The fact that upon opening the software, the user is introduced to a blank screen encourages the composer to think like a programmer, in terms of nonlinear and modular systems. Cascone details the emergence of a "nonlinear architecture" in the design shift from the "linear, tape deck kind of paradigm" offered by the *Pro Tools* plug-in interface to the visual programming environment of *Max*.[52] If procedural thinking is defined by the privileging of building from the smallest to the largest units—the opposite of thinking in terms of intentions, final structures, and end goals—then procedural listening in music production may be defined by an attention to the relationship between musical "modules" of various sorts. This systems-based approach is at the heart of learning computer programming and music-related programming, such as *Max*.

At the most basic level, the "incomplete" aspects of the *Max* interface are experienced through the discrete nature of its building blocks: objects. Whereas "traditional" music composition often involves the juxtaposition of precomposed musical ideas such as riffs, rhythmic patterns, melodies, and harmonies, the primary "content" of *Max* objects is not compositional ideas but algorithmic processes defined in relation to one another. These incomplete units "do nothing" on their own and can only be activated by connecting one to another through virtual patch cords. To construct a complete *Max* patch, the user must work through the interactions between objects in a flexible, step-by-step manner, aware of the multiple relational possibilities between objects at any given moment. Early computer graphics artist Manfred Mohr characterized this relationality between objects in automated systems as the paradox of generative art, as he claims, "formwise it is minimalist and contentwise it is maximalist."[53]

The step-by-step, algorithmic, and "incomplete" manner in which the patch is constructed fosters a diagnostic relationship between the composer and the software. As the patch "runs," the composer views the resulting program from a macro perspective, able to pinpoint the microprocesses happening every step of the way. In the random note generator, number boxes after the "random" and "makenote" objects provide composers with real-time feedback in the form of constantly changing integers, allowing them to perceive characteristics of the sounding note such as volume and duration. At the same time, the "bang" button blinks when triggered by the "metro" object, reassuring the composer that the patch is functioning correctly. In addition to teaching users to "think outside the timeline" of various DAWs, this process of debugging a patch in *Max* encourages the user to focus on the microprocesses of each individual object algorithm. The composer's understanding of the mechanics and rules of the software is regularly "updated" in response to the constantly changing "state" of the *Max* patch. Computational thinking in sound thus involves a feedback loop: the user diagnoses the status of the computational system while programming the software itself. In other words, this "integrative composer" is simultaneously creator, listener, producer, and consumer.

As a result of the diagnostic aspects of procedural listening in *Max*, composers employ rapid and iterative prototyping methods in the patch building process. Since the effects of any changes made to the patch are revealed immediately to the composers, they can substitute objects for one another in order to explore the possibilities afforded by the patch. In the current iteration of the random note generator, the "random" object functions to generate a random number which eventually determines the frequency of the resulting note. If the composer alters the "random" object and connects it to a "sel" object, the patch could randomly select notes among a specific set of options. For example, in the random triad generator (figure 3.14), the "random" object chooses among three different notes, which are determined by the "sel" object. The three possibilities are the three MIDI notes from the minor triad, as displayed in the "message" boxes below the "sel" object.

Among design professionals in the early twenty-first century, this type of iterative prototyping—defined by the constant *process* of testing, analyzing, and refining a product—has replaced more *fixed* text documentation as a core deliverable given to project managers.[54] *Swift*, Apple's programming environment for OS X and mobile software development, incorporates iterative methods in its workflow by introducing "playgrounds"—interactive documents where *Swift* code is compiled and run live as the user types.[55] Just

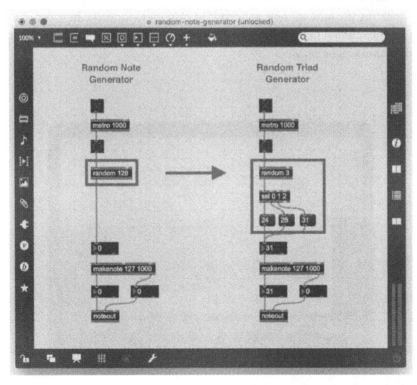

Figure 3.14 *Max* 7 random triad generator patch (screenshot taken by author).

as *Max* patches allow the composer a macro perspective on the operations of the patch, *Swift* playgrounds display the program's operations in a step-by-step timeline, allowing the programmer to inspect and revise the code at any point (figure 3.15). If code is often conceptualized as a text-based *object-oriented* representation of the *process-oriented* outcomes of software, the *Max* patch offers a tool for musical representation that's simultaneously a kind of musical score (an inscribed set of rules about sound) *and* performance (the mechanism through which those rules are sounded). The analogy to digital music composition is clear: in the context of procedural listening, people don't simply read musical notation; they also interact with it.

The ability to conceptualize both micro and macro structures of a *Max* patch is facilitated by an information architecture that arranges smaller units into increasingly larger structures: objects are arranged into subpatches, which are then connected to other subpatches to form a larger patch. Picture a more complicated random note generator, triggered not by manually clicking the "toggle" object in *Max* but by an external input such as the amount of light in a room. When the lights are on, the "toggle" is on, thus triggering the random note generator to play, and vice versa. In order to do this, the random note generator patch must be connected to a separate light-detection sensor patch.

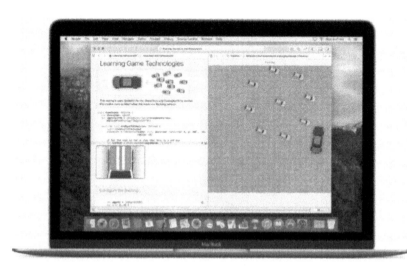

Figure 3.15 *Swift* "playground"
Source: Apple Developer, "Swift," accessed May 1, 2016, https://developer.apple.com/swift/.

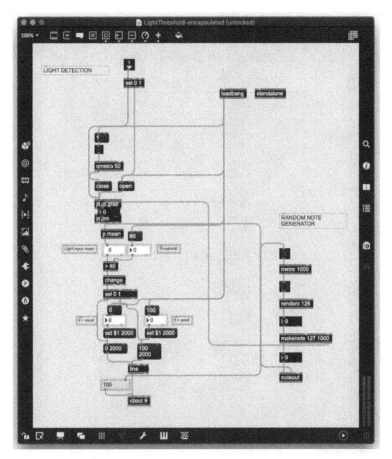

Figure 3.16 *Max* 8 light threshold patch (screenshot taken by author).

As figure 3.16 shows, the resulting patch is more complex visually than the random note generator by itself. Similar to other common programming languages, each patch can be "encapsulated" into a smaller subpatch (akin to a complex dropdown menu), thus clarifying the visual layout of the overall patch (figure 3.17). Encapsulation provides a simplified way to hierarchize functions within a patch, allowing the composer to follow more carefully the individual processes occurring within the larger program, as well as the relationship between the micro and the macro aspects of the patch design. The act of structuring musical processes on a hierarchical level is especially important when connecting patches within the *Max* interface to software outside of it. Ultimately, the supposed end product, the "complete"

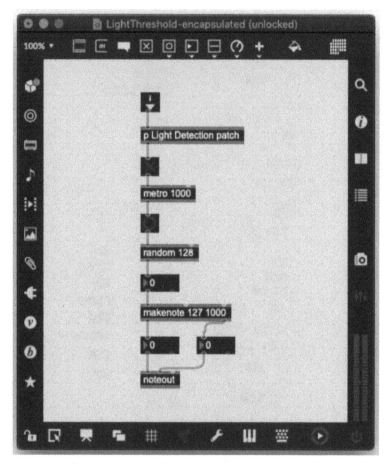

Figure 3.17 *Max* 8 light threshold patch encapsulated (screenshot taken by author).

patch, often serves as just another musical module within a performance environment that includes other musicians, instruments, or media platforms. The incomplete, modular, and iterative nature of procedural listening in *Max* encourages the application of systems-oriented design methods as a way of both handling algorithmic complexity and thinking of program functionality outside of a specific software.[56] ▶

Most *Max* objects are agnostic as to which media format is being controlled, allowing the integration of multiple media platforms and techniques into a single project. "Jitter" is a specific set of *Max* objects dedicated to video, graphics, and matrix data processing, allowing real-time cropping, stretching, and juxtaposition of moving images, as well as the handling of information through spreadsheets and databases. In turn, these text-based data-handling objects are capable of interacting with audio objects dedicated to synthesis, sampling, and editing. Further, *Max* is capable of interfacing with software and hardware outside of its native environment. By working with the software development kit (SDK) and the application programming interface (API), users can develop their own *Max* objects capable of sending control messages to web services such as YouTube, Spotify, and Last.fm, as well as hardware devices such as mobile phones, computers, and portable sensors of various sorts. When these integrated systems are set up, users can easily reroute audio to and from other music software on the same machine, using sound-routing applications such as *Soundflower* and *Rewire*. In the random note generator patch, for example, double-clicking the "noteout" object reveals a list of possible outputs for the resulting MIDI note, allowing the composer to send the output of the patch to other DAWs such as *Live* or *Pro Tools* (figure 3.18).

The integration of *Max* with external media platforms highlights another aspect of procedural listening: the ability to understand the techniques of "musical" and extra-musical software. In *Max*, the production environment is not simply the software alone but countless APIs and external objects that are meant to interface *Max* with other software applications and media platforms. Manovich describes this convergence of separate media techniques and tools into software as "deep remixability," a form of composition in which "the software production environment allows designers to remix not only the content of different media types, but also their fundamental techniques, working methods, and ways of representation and expression."[57] *Max* objects serve as abstractions of musical gestures (triggers, pitch bends, dynamic changes) and tools (metronomes, transport controls [play, pause, stop], beat counters), functioning similarly regardless of the actual content or platform that they control.

In the end, *Max* and *Pure Data* are programs that allow musicians to manipulate numbers. The software itself is agnostic to the output of these numbers, allowing a patch as simple as the random note generator to control actions ranging from the manipulation of interface elements in a video game to the thermostat levels in the composer's home. Cascone talks about developing a *Max* tool that enables sound designers to generate algorithms that can be used by game programmers: "They don't have to worry about talking the same language, because all he has to say is okay, I've exposed all

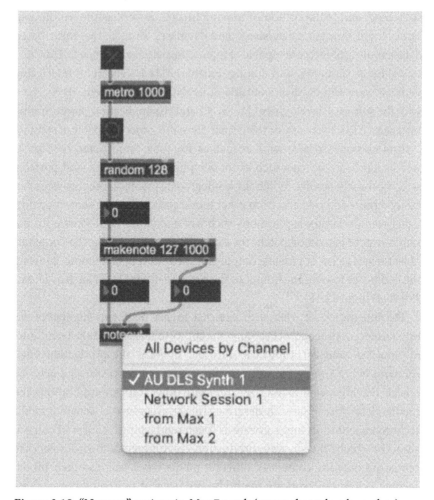

Figure 3.18 "Noteout" options in *Max 7* patch (screenshot taken by author).

these controls to you. You need to send values in this range with this name to these controls."[58] In facilitating a shared technical knowledge among media producers—what new media expert Noah Wardrip-Fruin terms the "operational logics" of technological systems[59]—*Max* encourages a shift from the isolated architecture of *recording* studios to the emerging *network* studio ideal, described in chapter 2. The integrated nature of this compositional model privileges a process-oriented relationality between media objects, transcending conventional musical parameters such as pitch, volume, and timbre.[60]

On the one hand, computational thinking provides fresh opportunities for educators and musicians alike to provide more inclusive spaces within what often seems to be the black-boxed world of computing. As Obama stated, what's important for education across the disciplines of engineering and the arts is that students are familiar "with not just how to play a video game, but how to create a video game."[61] This praxis-oriented "maker" model of digital literacy, in which users learn computational systems by building them, is fundamental to the experience of patching in the *Max* paradigm.

At the same time, the values and ideologies embedded within computational thinking and the *Max* paradigm—reason, logic, rationality, or what software designer and digital media theorist David Golumbia calls the "cultural logic of computation"[62]—have historically been used to exclude others from the creative and professional practices surrounding it. It's no coincidence that the majority of the developer and user base for the *Max* paradigm consists of white men. Indeed, not only does whiteness frame the terminal design of the blank white screen of *Max* and the racial demographic of its users, but the technological organization of knowledge at the heart of computer coding emerged at the same post–World War II historical moment as key racial organizing principles of a segregated America. Digital media expert Tara McPherson claims that this mode of social, cultural, technological organization is guided by a "lenticular logic," a logic of the fragment that sees the world as discrete modules or nodes, suppresses relation and context, and manages and controls complexity.[63] For McPherson, these principles are simultaneously embedded within the best practices of early computer programming languages (software developer Eric Raymond's rules of Unix programming, e.g.),[64] as well as a neoliberal multiculturalism that claims to be "post-racial." She describes how the contradictory push within free and open source software movements toward modularity, openness, and flexibility while also valuing covertness through encapsulation and hierarchical coding practices reflects a fundamental change in the organization of social life in the 1960s: "if the first half of the twentieth century laid bare its racial logics, from 'Whites Only' signage to the brutalities of lynching, the second half increasingly hides its racial 'kernel,' burying it below a shell of neoliberal pluralism."[65] In other words, as much as the free software movement values openness and freedom, the demographic makeup of live coding and many other electroacoustic communities invested in the *Max* paradigm tend to perpetuate straight male whiteness.

Think of the often hidden history of coders who were women and how the logic of computational thinking has been used to convert computer programming from an initially feminized practice to a "creative" and "professional" display of masculinity similar to the monopolies of competence detailed in chapter 2. During the advent of computer programming in the 1950s and '60s, women were the sole technicians to carry out these tasks, largely due to gender stereotypes at the time that claimed women were better suited to feminized clerical work such as coding and corporate data processing. Since personal computers hadn't been invented yet, this was also a time in which all students entering college computer science programs were entirely new to coding, thus placing everyone on an equal playing field. Journalist Clive Thompson details how the advent of personal computing in the 1970s and '80s helped to shift gendered perceptions of coding away from its initial associations with women.[66] To paraphrase a longer and more detailed history, personal computers allowed kids to code and play with computers before college. Gender dynamics and wealth gaps in the home and school often encouraged white boys, rather than girls and students of color, to pursue computing; boys were twice as likely to be given a computer as a gift than girls; if parents bought a computer for the family, they most often put it in a son's room; fathers developed "internship" relationships by coding with their sons and not their daughters; and geeky boys at school formed computer clubs in response to their exclusion from jock culture (clubs that became just as exclusionary against women). Media stereotypes in the 1980s and '90s further solidified the connotation of boys with science and computers through movies such as *Revenge of the Nerds, Weird Science, Tron*, and *WarGames*. Finally, as demand for professional coders increased with average salaries, the "capacity crisis" pushed women out of the field, since men were often the only coders who had the opportunity to develop programming skills prior to college.[67]

The birth of coding, post–civil rights America, the rise of neoliberalism, the advent of personal computing—all of these forces coalesced to form a single archetype that would reflect the interconnections between computation, whiteness, and masculinity in the twenty-first century: the geek. In the context of geek culture, computational and neoliberal principles of modularity, openness, flexibility, and rationality become more than just covert values embedded deep inside the technical structures of software design and development; they become a worldview in themselves. UCLA professor Christopher Kelty describes how, through coding, patching, sharing, compiling, and modifying software, geeks can both express and implement ideas about the social and moral order of society: "Geeks use technology as a kind of argument, for a specific kind of order: they argue *about* technology, but they also argue *through* it. They express ideas, but they also express *infrastructures* through which ideas can be expressed (and circulated) in new ways."[68] These praxis-oriented ideas reflect a vague libertarianism that privileges self-governance through a sort of invisible hand of the internet (in other words, if the code works, it has the possibility of surpassing human law). This ideology epitomizes the moral and technical order at the heart of the *Max* paradigm's terminal aesthetic, and it's a testament to Kelty's observation of free software communities that "the political lives of these folks have indeed mixed up operating systems and social systems in ways that are more than metaphorical."[69]

The image of the coder as social, cultural, and political outsider has helped to solidify further associations between the cult of genius in computer programming cultures and specific types of white masculinity within geek cultures. In a 1972 *Rolling Stone* article, tech visionary Stewart Brand positioned early hackers as an extension of the 1960s counterculture, describing them as "revolutionary," "outlaws," and "heretics" possessing the same kind of fanaticism as the greatest artists, inventors, and explorers.[70] Historian Nathan Ensmenger frames the obsession with technology, lack of conventional social skills, and inattention to physical health and appearance among male computer programmers as a crucial component in the ritualized displays of masculinity constantly being undertaken within geek culture and open source software communities. In university computer centers, for example, programmers engaged in marathon coding sessions, surviving for days on soda and junk food as a form of manly demonstration and competition. In this context, "The goal was not so much to accomplish an objective but to produce code that was beautiful, elegant, humorous, or otherwise aesthetically appealing. . . . The goal was simply to please oneself or, more frequently, to impress one's peers. . . . Trimming code served as a form of masculine competition, a means of both demonstrating mastery over the machine and establishing dominance within the community hierarchy."[71] Significantly, the earliest universities to develop computer centers, such as Princeton and Columbia, didn't allow women to enroll until the late 1960s. As Ensmenger notes, the sheltered and unsupervised environments of university computing centers thus became linked with the cultural practices of an adolescent masculinity that continues to persist within the corporate "campuses" of Google and countless other tech firms that include such amenities as "play areas," "nap rooms," and "tree houses."[72] Through their embrace of the cultural logic of computational thinking in sound, institutes such as IRCAM—and many other university research centers for electroacoustic music—have perpetuated this brand of straight white masculinity.

Open source software and computer programming are just a couple of outlets through which white men have mobilized masculinity as a way of achieving professional status and personal autonomy. The various character traits and belief systems expressed and embodied by many male coders reflect what historians Erika Lorraine Milam and Robert A. Nye term "scientific masculinity." Regardless of the scientific discipline, Milam and Nye note how male scientists often express their identities through a variety of masculine roles, including "laboratory-based scientist-heroes, outdoor, self-reliant men, sensitive and sympathetic readers of nature, and family men."[73] In each case, there's a dialectical relationship between the establishment of gender norms and professionalization processes, and spaces such as the lab, the classroom, and the field are mobilized to screen out women and others from full participation.[74]

This brief overview of the gendered and racialized history of computing in the United States at least partly explains why computational thinking has been so closely aligned with the professional and creative practices of white men and, as a result, why women continue to fight an uphill battle when it comes to real equality within not only electroacoustic music communities but the tech industry more broadly. As in many code-oriented communities, practitioners within the *Max* paradigm tend to be, for the most part, straight white guys.[75] I bring all of this up not simply to accuse any individual of being exclusionary toward women or people of color but rather to show how the very logics of social exclusion developed in post–World War II America emerged in parallel with neoliberal capitalism and its reliance on computational thinking. These gendered associations won't be solved simply by reversing numbers—hiring more diverse workers, shedding light on female participants within cultural communities. Like battles for equity in DAW design described in chapter 2, diversifying the cultural and technical aspects of the *Max* paradigm will require the building and sharing of social networks and technical resources. Steps in the right direction include Judy Dunaway's work with Max/MSP/Jitter Women Boston, the Diversity in Computer Music Scholarship for Programming Max/MSP offered by Stanford's Center for Computer Research in Music and Acoustics, and the excellent work of *Max* experts such as Yvette Janine Jackson, Margaret Schedel, and other women and BIPOC teachers of the software.

In these spaces, procedural listening may prove useful in shifting computation away from its emphasis on rationality, reason, and control and toward a type of relational HCI that negotiates between critical theory and practice. In turn, digital cultures themselves—from the free software movement to live coders and graphic artists—may evolve from their early-twenty-first-century role as "tinkerers" who produce technological *objects* into individuals capable of developing new *processes* of relating to society, culture, technology, and one another. Like so many of the programs I discuss in this book, *Max* and *Pure Data* leave us with the fundamental question: how might procedural listening allow us to hear the complex dynamics of social, cultural, and political interaction in the technical design of software itself?

PART II
WHEN SOFTWARE BECOMES HARDWARE

4

Controller Cultures

In the first decades of the twenty-first century, EDM and video games emerged as dominant forms of popular cultural expression. The rise of EDM as a global industry parallels the rise of massive multiplayer online video games, both demonstrating the power of social media to mobilize previously isolated communities of gamers and musicians. As a result of their widespread appeal, video games and EDM have become the targets of American politicians and moralists who blame the cultures for everything from drug use to gun violence and cultural decay.[1] In a similar manner to previous youth cultural movements, games and dance music simultaneously reify and subvert existing anxieties about the increased role of media and technology in everyday life and the resulting shifts in social relationships.

More recently, the visceral experiences of music and gameplay have converged, shaping the embodied practices of music and game creators. The success of music video games such as *Guitar Hero* and *Rock Band* has influenced both amateur and professional musicians to think through the practical connections between music performance and gameplay. Dubstep pioneers Benga and Skream have talked about how their use of the Sony PlayStation game console to make beats has shaped the sound of dance music.[2] In 2014, Red Bull Music Academy even launched a documentary series titled "Diggin' in the Carts," tracing the global influence of Japanese video game music from the 1980s and '90s on genres of electronic music.[3] Through interviews with game music composers and hip-hop DJs alike, the series reveals unexplored relationships between the now-pervasive experience of gameplay in everyday life and the creative practices of electronic musicians.

Combining theories of play with analyses of the technical design of both music and video game controllers, this chapter discusses the embodied practices of electronic music production in relation to the haptic control of gameplay. The concurrent rise of video games and EDM charts an alternative history in the evolution of digital media. Rather than reifying the centrality of "analog" technologies such as the turntable in the birth of popular music genres, the ongoing convergence of games and music establishes forms of

Push. Mike D'Errico, Oxford University Press. © Oxford University Press 2022.
DOI: 10.1093/oso/9780190943301.003.0005

experimental play with new media as crucial to the development of cultural production in the twenty-first century.

While software has become commonplace in the studio and on the stage, the history of hip-hop has always been rooted in the analog materiality and physical manipulation afforded by the vinyl record. Early DJs such as Grandmaster Flash, Grand Wizard Theodore, and Afrika Bambaataa established vinyl record mixing, cueing, beat matching, and beat juggling as techniques for aspiring turntablists, showcasing their individual virtuosity and new modes of performance made possible by technological experimentation. In his pioneering mix "Adventures of Grandmaster Flash on the Wheels of Steel," Flash popularizes the "cut"—a mixing technique in which the DJ quickly shifts the sonic output from one turntable to another—a gesture made possible by Flash's own invention of the crossfader. Bambaataa, on the other hand, was known for his eclectic musical tastes and extensive record collection. While disco records provided common musical fodder for early hip-hop DJs, Bambaataa experimented with rock, Latin, and other styles from around the world, including Kraftwerk's *Trans-Europe Express*. His 1982 track "Planet Rock" was an early prototype for the now pervasive concept of the remix, highlighting the turntable as a musical instrument and a symbol for the increasing circulation of culture through technology in the early 1980s.

Vinyl secured a key role in preserving hip-hop traditions by offering musicians a sense of tangible presence and real-time manipulation. Mark Katz, who has written extensively on the art and culture of the hip-hop DJ, claims the physical immediacy of the record is the most important reason for its success. He describes the hand resting "comfortably on the grooved, slightly tacky surface. . . . Pushing a record underneath a turntable needle, transforming the music held within its grooves, one has a sense of touching sound."[4] The "inimitable feel" of vinyl comes through not only in the performance practice of the DJ but also in the hands of record collectors who value the dusty, aged quality of vinyl just as a book collector values the original printing of a text. In manipulating the deep wax grooves on the surface of a record, DJs sense they are "touching sound" and are allowed immediate access to the musical source and social context embedded within the object. As such, vinyl sampling is important in asserting a lineage of black aesthetics through technologies of sonic reproduction. Sociologist Tricia Rose writes of sampling as a creative practice that "affirms black musical history and locates these 'past' sounds in the 'present' . . . these [sampled] artists have been placed in the foreground of black collective memory."[5] Further, Rose describes the "performative resonances" that result from the process of sampling, highlighting the aural continuity that exists between the sampled musician and the hip-hop producer.[6] Katz notes the "double-voicedness" implicit in these "digital form[s] of signifying," as the virtuosic sampling and looping of breakbeats serve to "draw upon and honor the work of the hip-hop DJ."[7] The playful manipulation of vinyl through "breaks" and "cuts" has functioned historically as a strategy for hip-hop musicians to reclaim control of black musical labor in increasingly deindustrialized urban spaces.

It's no coincidence, then, that an archaeological rhetoric pervades discussions surrounding record collecting within hip-hop. The act of seeking out new records for creative inspiration, known as "digging in the crates," has become a rite of passage for DJs. According to Joseph Schloss, an expert on hip-hop music production, "one of the highest compliments that can be given to a hip-hop producer is the phrase 'You can tell he digs.'"[8] Typical dig sites for these media archivists include yard sales, online auctions, the hidden recesses of thrift stores, and the basements of old record stores. ▶ The excavation of vinyl facilitates both the construction of hip-hop's musical genealogy and the expansion of technical innovation within the genre; Katz describes the materiality of the vinyl as "a precious substance in hip-hop" that is "authentic," "elemental," and "fundamental." Present at and largely responsible for the birth of hip-hop, Katz claims, "There is more than just music inscribed in those black discs; vinyl carries with it the whole history, the DNA, of hip-hop."[9] Similar to the mostly male practice of collecting carts of vintage video games, digging in the crates for vinyl records reflects a masculine bias in which the "subcultural capital" of both knowing and owning historical documents is used to legitimize one's sociocultural status.[10]

In the late 1990s through the early 2000s, vinyl culture would face a major challenge with the emergence of digital tools for music production. For a culture so intimately dedicated to the physicality of the record, what happens to the structure of hip-hop's musical DNA in the context of software's immateriality? How are techniques of production and performance coping with the perceived obsolescence of vinyl?

In 2010, Technics discontinued production of the SL-1200 turntable. The iconic model was lauded for its minimalist interface and direct drive system, which afforded the DJ a robust instrument with a clear sense of tactile feedback (figure 4.1). Countless obituaries surrounding the device's death marked this moment as the end of an era, questioning what would become

Figure 4.1 Technics SL-1200 turntable (1972)
Source: https://djworx.com/weekend-poll-silver-or-black-technics/.

of hip-hop in the post–SL-1200 age.[11] In the same year, Apple introduced the iPad, a touchscreen portable tablet that became popular among musicians seeking new ways of controlling the increasingly complex production software developed for laptops (figure 4.2). These coinciding developments turned out to have a major impact on the forms and techniques of hip-hop production and performance, marking the convergence of multiple spaces within EDM culture—studio producers become stage DJs, laptops converge with mobile devices, and software becomes hardware.

While turntablism thrives on the physical dexterity of the DJ and the visibility of the vinyl record, laptop musicians often struggle with creating convincing performances. Since the computer serves as the focal point of their stage setup, laptop DJs are often accused of playing video games or simply checking email without offering the audience an entertaining performance. DJ John Devecchis questions the idea of laptop performance as a form of DJing altogether, as he asks, "how do you know the DJ is even playing? How do you know he's not playing a pre-recorded set? How do you know he's not playing Pac-Man while he's supposed to be DJing? I want to see the DJ doing something."[12] For Devecchis, other DJs, and fans of EDM, the lack of visibility in performance techniques delegitimizes the skill of the performer.[13]

Debates concerning legitimate techniques of electronic music performance emerged on the heels of such technological changes, eventually

Figure 4.2 Apple iPad (2010)

Source: https://www.esquireme.com/content/43065-happy-birthday-ipad-apples-tablet-turns-10.

coming to a head in 2013 as a result of a controversial statement by Joel Zimmerman, also known as Deadmau5, one of the most globally renowned DJs at the time. In a blog post titled "We All Hit Play," Zimmerman claimed to speak for all the "button pushers" who were too afraid to admit that most DJs' "live" performances consist of simply getting onstage and pressing play: "its [*sic*] no secret. when it comes to 'live' performance of EDM . . . that's about the most it seems you can do anyway. It's not about performance art, its [*sic*] not about talent either (really its [*sic*] not)."[14] In direct response to DJs such as John Devecchis, who prioritizes individual skill and "paying your dues" as a turntable DJ, Zimmerman celebrates the *lack* of skill and technical accessibility of DJing in the digital age, claiming that "given about 1 hour of instruction, anyone with minimal knowledge of ableton and music tech in general could DO what im doing at a deadmau5 concert." The post immediately went viral among the online community of DJs, inspiring heated exchanges and countless defenses of the lineage of live performance in DJ culture, including Twitter rebuttals from Zimmerman's friend and fellow DJ Sonny Moore, also known as Skrillex.

The "button pusher" debate highlights many of the ongoing anxieties musical cultures experience with the rise of new technologies. For some audience members, the presence of a laptop onstage negates the live aspect of the event and thus their own physical presence at the club, leading them to think, "Why not just listen to the music in the isolation of my home?"[15] For some DJs, particularly those who have dedicated years of their lives to learning turntablism techniques, the laptop delegitimizes the creative labors of a musical tradition nearly half a century old. These moments of discursive tension have always arisen during periods of technological innovation within dance music culture, reflecting the nature of technological change in society more broadly.[16]

In gauging the quality of a DJ performance based solely on a particular group of skills developed by the *human* musician, the "button pusher" discussion is limited in its focus on the individual performer's creativity. As such, it ignores the media environment within which DJ performances are embedded, including the influence of stage lighting, visual displays, and constantly changing technologies of musical performance. The level of musical control afforded by the human body, while privileged by many producers and audience members, is just one of moving parts that come together to form the electronic music event. In defense of Deadmau5, one of the more significant takeaways from the situation described in "We All Hit Play" is Zimmerman's attempt to emphasize the importance of the seemingly extra-musical factors such as multimedia accompaniment and audience energy levels in contributing to the overall "performance."

Especially during these moments of discursive tension in dance music culture, the setup of instruments and other musical gear plays a significant role in constituting the performance environment. Rather than viewing the technologies as threats to performance traditions, as in the "button pusher" debate, music theorist Mark Butler describes the increasing prevalence of hardware "controllers" in the laptop performer's setup as tools for making transparent the seemingly opaque creative processes happening behind the laptop screen. According to Butler, "Rarely if ever is a 'laptop set' *only* a laptop set. Instead, the internal, digital elements of the laptop environment are externalized—made physical in the form of MIDI controllers and other hardware devices."[17] In the wake of Zimmerman's critique of the state of performance in dance music culture, both DJs and producers have increasingly turned to hardware controllers as a means of heightening the physicality and theatricality of their live presence.

"Controllerism" emerged in the late 2000s against the heated backdrop of the "button pusher" debate. While the term is used to describe several performance techniques in electronic music, musician and hardware hacker Moldover defines it as being "about making music with new technology. Right now controllers are where it's at, and so that's the name for the movement. Button-pushers, finger drummers, digital DJs, live loopers, augmented instrumentalists; we're all controllerists."[18] For Moldover, controllerism reflects a unique stage in the development of music technology, one that arose at a historical moment in which the vinyl record ceased being the only interface for performing prerecorded musical material. The spread of digital music controllers has defined electronic music production amid the perceived twilight of vinyl, helping DJs and producers to navigate emerging tools and techniques through new forms of musical practice.

The racial dynamic of controllerism's apparent break from vinyl records is significant. While the demographic makeup of electronic musicians practicing controllerism is diverse, in terms of both gender and race, Moldover's comments reflect a conception of technological creativity that aligns with a model of white privilege practiced by similar tech communities such as the free software movement, hacker culture (see chapter 3), and gamer culture (see chapter 6). Whereas creatively repurposing music technology by young black men is often perceived as being tied to a sociopolitical agenda of resistance, tinkering with technology by white men is often praised as a form of apolitical "play." Ryan Diduck illustrates this distinction by comparing how samplers were portrayed in *Ferris Bueller's Day Off* and *The Cosby Show*. In *Ferris Bueller*, the sampler was "yet another in a collection of bourgeois fad gadgets . . . used to produce leisure by circumventing labor." On *The Cosby Show*, it was "constructed as a central technology to eliciting the uncanny . . . as well as being a machine of empowerment, with potential for economic and creative benefit."[19]

Further, the self-proclaimed archival work being done by vinyl-oriented hip-hop DJs is often tied to cultural and political movements centered in black communities, while discourse surrounding controllerism is rooted in internet communities such as music gear web forums and music tech blogs. This dichotomy echoes literary critic Samuel R. Delany's distinction between what he calls the "white boxes of computer technology" in software-based hacker culture and the "black boxes of modern street technology" in hip-hop's reclamation and resampling of sonic hardware.[20] Riffing on Delany's idea, Alexander Weheliye—scholar of black literature and culture—claims that technological play might be able to function as a platform for alternative storytelling and enunciation, outside of the dominant frame of whiteness in technology communities.[21] Controllerism is best understood, then, as a recent example within a longer history of black aesthetics in popular music production—from the experimentalism of Jamaican dub reggae producers like Lee "Scratch" Perry to the technological remixes of Angolan Kuduro producers—of musical cultures that exist between "resistance" and "reification" narratives of technology. These musical communities epitomize what Paul Gilroy calls "counter-culture[s] of modernity, forged through the diametric but symbiotic opposition" between black aesthetics and the tech industry.[22] Like so much of the music that emerged from the African diaspora, controllerism's cultural politics were born out of a resistance to American capitalism and consumerism as well as a strong embrace and reliance on the tools of global capitalism in order to carry out its politics.

In terms of controllerism techniques, the use of MIDI devices to control software is the most common practice, with historical roots in the rise of digital technology in the 1980s. ⊙ Created in 1983, MIDI was an important development in the history of music technology in that it facilitated increasing connections between "digital" computers and "analog" hardware such as synthesizers.[23] The initial specifications were limited in terms of what types of connections could be made between instruments—allowing one synthesizer to control another synthesizer's notes or output volume, for example. However, after a few years, MIDI features were adapted to many early computer platforms, including Apple Macintosh, Commodore 64, and Atari ST.

In popular music since the early 2000s, MIDI devices have not been used simply to send control messages between instruments but also as live instruments to be manipulated in real time. Grid-based interfaces with rubber pads have become commonplace in the studio and on the stage, allowing the percussive triggering and automated sequencing of digital samples. Ableton's APC, Livid's OHM, the Monome, and Novation's Launchpad, among many others, specifically cater to the live triggering and micro-manipulation of both musical patterns and sonic parameters such as volume, effects, and mixer settings (figure 4.3). Other grid controllers are fashioned as entire studio workstations in themselves. Native Instruments

ABLETON PUSH 2   AKAI MPC X   NATIVE INSTRUMENTS
                                        MASCHINE+

AKAI FIRE   AKAI MPK MINI

Figure 4.3 Grid-based MIDI controllers (screenshot taken by author).

describes its Maschine Studio as an "ultimate studio centerpiece for modern music production," specifically emphasizing the "unprecedented physical control and visual feedback" of the interface.[24] Designed by Ableton in collaboration with Akai Professional, the company responsible for the infamous MPC series drum samplers, the "Push" controller is likewise marketed as a digital controller that blurs the line between production and performance, presenting a comprehensive degree of fine-tuned control while composing using Ableton *Live* software. ⊘

While grid-based controllers dominate the digital instrument industry through a carefully marketed alignment with proprietary music software, other controllerists feel limited by the creative constraints resulting from this integration. Brian Crabtree and Kelli Cain started building open source, minimalist controllers in 2006, seeking to construct "less complex, more versatile tools" than the cluttered interfaces being marketed to electronic musicians at the time. Their company prides itself on operating "on a human scale," using only local suppliers and manufacturers, and embodying values of environmental sustainability in the design process. This minimalist sensibility is embedded within products such as the Monome "grid" controller, in which the only control mechanisms on the instrument are small rubber buttons that send on and off messages to software such as *Max* (figure 4.4). ⊘

Figure 4.4 Monome "grid" controller (2015)
Source: Monome, "grid," accessed August 27, 2020, https://monome.org/docs/grid/.

Rather than perform with the seemingly prescribed options of proprietary software, Monome users build and freely share software patches that can be applied across a variety of artistic genres and creative needs.

As these examples demonstrate, controllerism surfaced as an attempt by electronic musicians and designers to employ hardware as physical extensions of existing instruments, simultaneously enhancing the sense of tactile immediacy imbued by turntablism and distinguishing themselves from the "button pushers" detailed by Deadmau5. Indeed, Moldover defines the main goal of controllerism using the same critical language as vinyl purists, claiming "performers who use computer technologies as musical instruments needed a way to differentiate themselves from people who 'check their e-mail.'"[25] In developing more integrated HCI systems, controllerists negotiate shifting conceptions of human and technological agency, embodied performance literacies, and computational forms of instrumentality. How might the stakes surrounding live electronic music in the 2000s simultaneously challenge and reify existing concepts of musical performance and instrumental embodiment?

Controllerism represents one example in a longer tradition of electronic musicians learning to navigate changing technologies. The question of how to perform electronic music in a live setting has been asked since the advent of tape music and electroacoustic traditions in the mid-twentieth century. When Pierre Schaeffer developed the concept of "concrete music" in the late 1940s, he was concerned with how new forms of composition with phonograph discs could liberate listeners, allowing them to hear sounds acousmatically—separate from their source, no longer dependent on "preconceived sound abstractions," and removed from the "elementary definitions of music theory."[26] Foreshadowing concerns surrounding the visibility of electronic music production techniques from the audience's perspective, Schaeffer quotes philosopher Paul Valéry: "Looking at this seashell, in which I seem to see evidence of 'construction' and, as it were, the work of a hand not operating by 'chance,' I wonder: Who made it? . . . *And now I strive to find out how we know that a given object is or not made by a man*."[27] This attempt to discern whether a human agent constructed a given sound lies at the heart of the acousmatic dilemma in electronic music, from tape and electroacoustic music to contemporary "button pushing."

The question of what makes a mediated performance "live" has been fundamental to the work of a range of thinkers and performers across the arts. Performance theorist Peggy Phelan defines performance as "representation without reproduction," arguing that "performance's only life is in the present" and that it cannot be "saved, recorded, documented, or otherwise participate in the circulation of representations *of* representations." In other words, if a performance was to be recorded in any way, it would become "something other than performance," since performance "becomes itself through disappearance."[28] For Phelan, the essence of performance lies in the temporal uniqueness of a real-time event opposed to mediated representations. This ontology aligns itself with a politics of identity that is always in opposition to constructed representations of subjectivity through media formats. Performance, in this context, is a channel through which subjects can escape the confines of hegemonic forms of mediated representation.

Those interested in music performance have extended this presence-based mode of thought, valuing the individual performer as the spotlight of the creative act. Expanding on existing ideas about the ineffability of music performance—existing in time as a material acoustic phenomenon—musicologist Carolyn Abbate suggests that "real music" is most clearly expressed in "an actual live performance (and not a recording, even of a live performance)."[29] Ethnomusicologist Charles Keil developed the notion of "participatory discrepancies" (PD) to highlight how performers, in their face-to-face relationships onstage, create differential tension in the expected rhythmic, tonal, and textural aspects of precomposed or otherwise patterned musical processes. Similar to Abbate's notion of "real music," Keil's PD model is rooted in the material, physical, and corporeal aspects of performance: sticks tapping metal, fingers plucking strings, a "dialectical materialism in action."[30] Keil belittles electronic music for being out of touch with the "natural" world: "The expanding mix of 'mediated and live' musics seems like a limbo or purgatory to me because the organic feedback loops are not complete and co-evolving, not in touch with nature and neighbors, always limited in at least a few ways by electricity, machinery and commodity forms."[31] Employing a similar rhetoric to Phelan's, Keil creates a split between the ephemerality of the live event and the implied freedom of the human performer on one hand and the rigid representational structures of electronic media and postindustrial capitalism on the other.

Questioning the privileging of the corporeal among intellectuals, media theorist Philip Auslander describes how impressions of "liveness" arise from the influence of technology and digital media in a performance context. In direct opposition to Phelan, Auslander suggests that the historical trajectory of artistic performance in the twentieth century is defined by the rise of mediatization as a core component of what's perceived to be "live" performance. He writes, "Initially, mediated events were modeled on live ones. The subsequent cultural dominance of mediatization has had the ironic result that live events now frequently are modeled on the very mediatized representations that once took the self-same live events as their models."[32] Examples of the desire for recorded musical aesthetics in live performance include the use of autotune onstage, live video performances at EDM shows, and the emergence of posthumous duets.[33]

While troubling the binary between "live" and "mediated" performance, positing the term "liveness" functions to retain the primacy of the artists' *presence* and *agency* as the measure of performance. This is echoed in the language of electronic musicians themselves. Composer Simon Emmerson details how changes in software have facilitated a shift in electronic music performance from the studio to the stage. For example, the advent of "live coding," introduced in chapter 3, highlights how "what used to be the studio's domain becomes available for 'live' working."[34] The emphasis on "real-time" computational processes foregrounds the presence of the "analog" human body interacting with the "digital" computer. For Emmerson, the blurred line between creative practices in the studio and on the stage reflects what he calls a "reanimation" of technical objects previously thought to exist outside of what's traditionally been considered musical performance.[35] In contrast to the more outdated anxiety that humans may one day disappear into a sort of machine-dominated cyber-reality, Emmerson reasserts the centrality of the body in HCI, claiming the agency of human presence as the source for a "reanimation" of "dead" technologies that lack the capabilities for dynamic change in performance.

Electronic musician Primus Luta presents a similar conflation of human agency and "real-time" performance by outlining what he calls "variability" in live electronic music performance—referring to how music technology affords real-time manipulation by both human agents and other technical objects.[36] In this context, assessing the liveness of a given electronic music performance is made possible by understanding the potential variability of a particular instrument *in relation to* how a performer is or isn't exploring that variability. However, Luta explicitly states that his model is meant to describe the performance of electronic music in a way that makes it "easier to parallel with traditional western forms," and his analogies to jazz make it clear that "variability" in this context is understood as an extension of virtuosic improvisation.[37] Liveness is once again conflated with the physical processes of sonic manipulation afforded by the human body in resistance to technology.

While many DJs and producers see clear links between their own work and the work of jazz musicians, they downplay the unique aspects of HCI in dance music performance by privileging values such as human agency and virtuosity. In a response to Luta, Robin James argues that *artist* agency in performance remains the underlying concept and value driving Luta's project. Since Luta works from the widely shared perspective that "electronic music is the new jazz," he ends up transposing jazz aesthetics "into terms compatible with electronic instruments and genres," according to James.[38] In an attempt to push the discussion beyond the frame of modernist aesthetics, James suggests that we think of musical instruments themselves as performers. This concept of decentering the "human" from understandings of performance allows us to think of musical production as an environment, or network, of reciprocal feedback between human and computer.

Eliot Bates, who has written extensively on music production, takes the notion of agency even further, thinking through how musical instruments can be regarded as causes of social interactions, rather than incidental effects of those relationships. Bates defines instruments as social *actors*, neither subjects nor objects but sources of action that imply, quoting philosopher

Bruno Latour, "no special motivation of human individual actors, nor of humans in general."[39] Covering a wide variety of musical genres and cultures, Bates outlines how instruments take on lives of their own, possessing "a propensity to teach their owners how to play them," channeling spiritual powers, and mediating interpersonal disputes in communities.[40] In the context of musical performance, these ideas draw attention to the instrument as an object that shapes and is shaped by the performer, encouraging a more relational view of the creative process.

Within EDM, musical recordings and digital controllers are two of the primary "actors" involved in the broader "network" of performance. Butler suggests that recordings create dialogic "networks" between *processes* of musical production and their material *products*, thus providing solutions to the perceived split between absence and presence in performance. Attributing equal ontological significance to recording, performance, composition, and improvisation reveals the dynamic interaction of these practices constantly at play within both studio production and stage performance.[41]

In the next section of this chapter, I propose a play-oriented model of electronic music performance. Rather than focusing on either the live agency of the human performer or the processes of sound manipulation afforded by the technological hardware, a play-oriented model of performance acknowledges the experimental negotiations that continuously emerge in every HCI. Encapsulating the core elements of procedural listening introduced in chapter 3, play provides digital media producers with a necessary skill set for understanding the "rules" embedded within software. While video games emanated at a similar historical juncture to that of the musical genres under discussion, the rise of controllerism foregrounds the significance of embodied experimentalism inherent in what media theorists have termed the "ludic turn" in contemporary culture.[42] In addition to voicing metaphorical connections between the materiality of video games and music production, electronic music producers have increasingly rooted their creative techniques and practices in the playful logic of gaming.

If vinyl record performance foregrounds the agency and presence of the musician, controllerist performance foregrounds the negotiation between the musician and the rules of the software. This relationship between hardware (human bodies, material technologies) and software (processes, logics, and mechanics of code) is analogous to the structures of video game play. "Button pusher" is not a denigrating term for artists working with hardware controllers but a metaphor for the convergence of a gaming logic with music production. Speaking of his own influences from gaming, Flying Lotus talks about growing up as an only child who "didn't have too many friends, but I had Nintendo." Like many electronic musicians growing up in the 1980s, the dawn of the gaming age, FlyLo cites that period as formative in his creative development, proudly stating that "Those sounds are part of my youth, part of my history."[43] Glasgow's bass music pioneer Rustie talks about how his production styles emulate the way gamers play, describing his experiences with the electric guitar and video games as "different means to the same end, really . . . there's not much difference between plucking a string and pressing a button, I think."[44] The 2000s witnessed the emergence of a new generation of electronic musicians, who grew up on Nintendos, Game Boys, and Ataris rather than their parents' record collections, and the performance practice of pressing buttons and swiping screens reflects this.

The term "ludomusicology" was introduced in the 2010s as a model to analyze the shared experiences of play, performance, and digital embodiment in gaming and music production. According to music and gaming expert Roger Moseley, ludomusicology is concerned with "the extent to which music might be understood *as* a game"—as a system of rule-based logics that "constitute a set of cognitive, technological, and social affordances for behaving in certain ways, for playing in and with the world through the medium of sound and its representations."[45] If, as Moseley suggests, musical scores, software code, and hardware interfaces constitute "the ludic rules according to which music is to be played," what might the technical practices of digital music producers say about the shifting nature of musical performance and instrumentality *as* play?

Thinking about constraints as engines for creativity within interactive systems is one way of explaining the allure of play as a cultural force.[46] In a definition that could be applied equally to music and gameplay, game theorist Bernard Suits describes gaming as "the voluntary attempt to overcome unnecessary obstacles."[47] Whereas musical play is often conceived as allowing an unfettered creative experience—the idea that technologies allow for the creation of "any sound you can imagine"—embodied interaction in games and electronic music may be more aptly characterized by how the media *resists* or *constrains* the actions of the user.[48]

Game designer Brian Upton details how constraints operate across four categories of a player's experience: technical design ("the game as designed"), player experience ("the game as encountered"), sociocultural understandings and real-world knowledge ("the game as understood"), and practical experience ("the conceptual background").[49] Upton's model aligns technical and cognitive aspects of creativity with social and cultural experience, emphasizing that although constraints are *embedded* within the design of a game, it's through play that they become *embodied* values in the player. Game design—and the creation of "playful" systems more broadly—is less about building one-way systems that respond to direct player input and more about encouraging the formation of internal constraints that facilitate creativity in the mind of the player. In the context of music, these internal constraints constitute years of musical training. Similarly, games construct these constraints through rote repetition in the players' experience of gameplay itself.

Whether embedded within music making or gameplay, constraints are often perceived in the physical comportment of players as they interact with a technological apparatus, the interface shaping their embodied knowledge and practices. Thinking through how the bodies of mobile media users "curve into supportive architectures with which they cradled touch-screens," dance expert Harmony Bench suggests that these "digital media choreographies" introduce new understandings of physical comportment and serve as the mechanisms for that education.[50] Bench aligns musicianship with the "computational literacy" of gaming, explaining the significance of rote repetition in the development of embodied knowledge within each practice, as well as how each "demand[s] a corporeal training that impacts operators' experiences of their physicality."[51] Think of how musicians, gamers, and computer operators alike constantly update their skills based on the rapid changes made to operating systems, game controllers, and musical interface design.[52] Bench's analysis is not limited to a single platform, allowing her to highlight gaming and music production as shared avenues for the embodiment of systematic design constraints that shape the bodily comportment of the player.[53]

Game controllers are important conduits for the transmission and negotiation of design constraints, aiding in the embodied cognition of social values, haptic metaphors for technological interaction, and expected patterns of use. In other words, controllers externalize the rules embedded within digital systems. According to game theorist David Myers, all video game controllers share at least two formal properties that directly shape players' embodied practices: "they employ arbitrary and simplified abstractions of the physical actions they reference, and they require some level of habituation of response."[54] While early video game controller *hardware* was often designed to match the on-screen actions afforded by individual game *software*, contemporary controllers have adapted more uniform design conventions, facilitating a steeper learning curve while allowing the device to function across a variety of games. For example, PlayStation 5 and Xbox Series X controllers (the most popular controllers at the time of writing) are similar in their dual-joystick layout, abstracting a complex set of buttons and triggers to letters and shapes (figure 4.5).[55] The fact that the buttons on the controller are mapped to the representational actions on-screen in an abstract way encourages the player to focus on what Myers calls "locomotor play," a form of technological engagement specifically involving "the manipulation of the interface between our bodies and our environment."[56] Abstraction in hardware is thus used to manage the complexity of software, allowing the player to physically internalize the constraints of the controller that are required to succeed in a variety of gaming genres. How might these design constraints apply to digital music making, a practice that asks the musician to navigate complexities in HCI and the conflicting perspectives on liveness previously discussed?

Like the development of motor memory in video games, musical training involves an internalization of the affordances and constraints of a given instrument *through* the rote repetition of bodily techniques and habituated responses. Elisabeth Le Guin discusses how cellists physically comport themselves in relation to the cello during performance, molding themselves into a single "cellist-body" through movement and action.[57] Just as gamers embody the internal constraints embedded within the game itself, instrumentalists develop an embodied understanding of the constraints embedded within a given piece of music. Le Guin defines this skill as "anticipatory kinesthesia,"

Figure 4.5 PlayStation 5 (left) and Xbox Series X (right) controllers (http://www.pngall.com/playstation-5-png/download/47893; https://www.citypng.com/photo/1744/front-view-microsoft-xbox-series-x-controller).

in which the performer assesses the physical demands of a given piece on their body, asking, "What do I need to do in order to play this? Where will I put my hands, and how will I move them?" Most instrumentalists couldn't articulate clear answers to these questions, in the same way that most gamers would have trouble putting into words such a deeply embodied practice. Rote repetition is thus capable of facilitating the acquisition of tacit, embodied knowledge.[58]

Electronic musical instruments similarly function through arbitrary and simplified abstractions of physical actions. While instruments such as a violin, an acoustic guitar, or a snare drum reflect a 1:1 ratio between the physical gesture of instrumental attack (bowed, plucked, or strummed strings, e.g.) and the sonic output (string vibrates at a specific frequency), digital music controllers can be "mapped" to any number of sounds. Tapping a pad on a drum machine connected to Ableton's *Live* could just as likely trigger a single snare drum sample, or an entire multimovement symphony. In the same way that game controllers externalize the rules designed with the game, the constraints of electronic music production and performance are found in the limitations of the instrument's formal structures, rather than the audio content being created. In confining oneself to a simplified, abstract *hardware* interface (a sixteen-pad grid of rubber buttons, e.g.), the performer can creatively exceed what's often perceived to be overdetermined and complex music production *software*.

As is the case with embodied, tacit knowledge in game controllers, the arbitrary mappings of musical software onto hardware ask the player to internalize a changing set of embodied musical techniques. This process of interface abstraction is exemplified in the minimalist design of the Monome "grid" controller, which is a small rectangular box fitted with a symmetrical grid of small rubber buttons and a USB port.[59] Often, the Monome is used as a controller for the *Max* visual programming environment, itself a modular, open software used for a variety of creative practices from electronic music synthesis to the real-time generation of 3D visuals. In this context, the button grid interface can take the form of a pitch controller alternative to the keyboard interface, an externalization of a step sequencer, a multitrack mixer or effects modulator, a visual spatialization map, and any number of other tools. Approaching the blank, terminal interface of an instrument such as the Monome, the musician must focus more on the internalization of specific software affordances than on the external affordances of the minimalist hardware.

This internalization of software (what we might call "music as designed," following Upton) *through* hardware ("music as encountered") has two seemingly opposing effects on electronic music production. First, as the processing power for a given musical task is increasingly delegated to the software, the physical and gestural manipulation of the hardware becomes increasingly unnecessary. This fact is highlighted by trends in game controller and interface design more broadly, which value the least amount of effort to achieve the maximum output. In the context of games, a single, slight flick of a PlayStation 5 controller's right trigger may just as likely fire a gun, swing a sword, open a door, or detonate a series of explosives. In the context of music, the single tap of a rubber pad may just as likely trigger a single snare drum sample, a four-bar drum loop, or an entire album. In valuing the arbitrary design of musical gestures, digital music controllers encourage musicians and audience members to develop new forms of embodied listening and production. It's this transitional moment that sparked the vehement and ongoing debates about human agency in performance detailed in the opening of this chapter.

Increased complexity in software design seems to facilitate a decreased complexity in hardware design, leading to what game theorist Bart Simon terms a "gestural minimalism" in gaming that could equally apply to musical performance. However, as players develop an embodied knowledge of the software's rules, they're able to dedicate more attention to the physical control of the hardware itself. This leads to the common experience of what Simon calls "gestural excess" in gaming, when physical movements are made in excess of what the hardware can perform.[60] For example, even though the joystick of a controller may be the only mechanism capable of steering a car in a racing game, the player often exceeds this limitation by gesturing with the controller itself *as* a steering wheel, dynamically contorting the entire body to the left and right as if controlling an actual car. This becomes a subconscious attempt to overcome the arbitrariness of the digital "mapping" by foregrounding the embodied metaphor on which the software is designed. Just as these gestures function to translate the rules of the game to the player, embodied metaphors can likewise translate a sense of musicality and performativity to an audience.

For electronic musicians, gestural excess represents a clear strategy for conveying a sense of liveness to their audience, while developing performance strategies for the embodied control of musical techniques embedded in software. Describing a performance from German electronic musician

Stefan Betke (also known as Pole), Butler writes about what he calls the "passion of the knob," in which the producer "seems to put his whole body into the extended turning of a knob," directing an "exceptionally intense expressivity toward a small, technical component associated with sound engineering."[61] These gestural excesses are highly choreographed, as the performer "telegraphs 'expressivity'" to the prerecorded musical material, locating themselves as the primary agent of the sounds being heard by the audience.[62] In a way, this mode of performance is meant to foreground the "human" presence while effacing the technological apparatus. At the same time, highlighting the physical practice of interfacial mediation likewise foregrounds the mechanics and rules embedded within the apparatus, thus indoctrinating the audience into new modes of listening to the interface. In other words, gestural excess gives the audience a practical method for listening to the electronic music controller as a *process*-based musical instrument, rather than a tool solely used in the composition of sound *content*. ⏵

Producer and DJ Daedelus has become infamous for his use of controllers to externalize the mechanics of music software in performance. The relationship between gameplay and music is further highlighted by the type of creative work to which he dedicates himself, including interactive audio installations, sound design for video games, and controllerism in live performance. In a fitting video shoot produced by the news and media website Into the Woods, he performs an entire "DJ" set in the middle of Portland, Oregon's Ground Kontrol arcade.[63] The video begins with Daedelus challenging a fellow beatmaker to a game of *Street Fighter* 2, followed by a montage of clicking and clacking button presses that trigger short bursts and choppy audio samples from the machine. Surrounded by the flashing lights, bleeps, and blips of vintage game consoles, the gestural excess of these two button pushers transitions seamlessly into Daedelus's musical performance (figure 4.6).

As the camera shifts focus from the game consoles to the musician standing in the middle of the arcade, the visual frame foregrounds a technical setup consisting of a laptop and two Monome controllers. The "brain" of the operation is the MLRv *Max* patch (discussed in chapter 3), which allows Daedelus to control simultaneously the playback and fine-tuned editing of musical parameters in multiple audio samples. The GUI consists of eight horizontal rows, each containing a sample, with options to adjust volume, playback speed, and pitch just below each row (figure 4.7). Using the Monome grids as controllers for MLRv, Daedelus then physically manipulates the rows

Figure 4.6 Daedelus performs at Ground Kontrol arcade, Portland, Oregon (2012)

Source: ITWTV, "Far From Home #9Daedelus Live Arcade Set," uploaded July 9, 2012, accessed August 27, 2020, https://www.youtube.com/watch?v=QY3mHBj_dhs.

Figure 4.7 Monome 256 grid overview during Daedelus's performance

Source: ITWTV, "Far From Home #9Daedelus Live Arcade Set," uploaded July 9, 2012, accessed August 27, 2020, https://www.youtube.com/watch?v=QY3mHBj_dhs.

of audio in various ways. The 256-button grid serves as the primary control mechanism, mirroring the layout of MLRv by dividing the 256-button
grid into sixteen rows. The rows then spatially fragment the corresponding
audio sample into sixteen parts, allowing the musician to play back specific
moments in the sample by pushing the buttons within the horizontal row.
The audio waveform in the software literally becomes externalized in the
hardware, and the rules of MLRv become playable. ▶

Daedelus's performance mannerisms further highlight the gestural excess witnessed during the gameplay depicted at the beginning of the video.
The Monome grid is angled upward, away from the performer and toward
the audience, and the laptop screen is out of sight, highlighting the physical
interaction between the musician and the hardware. Every button press is
accented by a rapid withdrawal of his hand from the interface, spatially exag-
gerating the spectral morphologies of the sounds being controlled. While the
256-button Monome remains stationary, Daedelus twists and contorts the
smaller sixty-four-button Monome, controlling audio effects mapped to the
device's accelerometer (the same sensor used in mobile phone technology).
Rather than simply "pressing play" and letting the computer do all the work,
these moments of gestural excess—combined with the abstract and minimal
design of the hardware device—allow the viewer to focus visually and aurally
on the musical patterns as they're chopped, stuttered, and looped in real time.

The video brings to the fore key elements of the shifting nature of HCI detailed throughout this chapter. First, the virtuosic performance practices employed by Daedelus using grid-based controllers highlight how new media
remediate the technical practices embedded within previous technologies (in
this case, the controller remediates the turntable). Second, using *hardware*
in performance to foreground the processes at play within music *software*
demonstrates an increasing balance between object- and process-oriented
perspectives on performance. Finally, the explicit alignment throughout the
video of musical culture with gaming culture exemplifies the shifting media
genealogies in hip-hop and EDM culture outlined at the opening of this
chapter. Not only is Daedelus remixing musical *content*, but he's also playing
with the *forms* of relation between musical hardware (bodies, technologies)
and software (algorithmic processes, rules).

As mentioned previously in the context of game controllers, the process of developing *embodied* instrumental technique with electronic
music controllers consists of two steps: internalizing the affordances and
constraints of the music software and externalizing those design mechanics

in the hardware. By foregrounding the hardware over the software, the complexities of the software processes can be channeled through a material device, and the feeling of non-mediation and "direct manipulation" results. As electronic musician Ander claims, the "most important thing I wanted to do was to get rid of the screen. I don't want to have a laptop on stage." Like many other producers and DJs, Ander uses a controller because it "gives you much more direct access to the music as well as to the audience," the minimal and abstract mapping capabilities offering "something which is easy to look at, where you can get a lot of information in a short time." As with Daedelus's use of the Monome to externalize and make visible the functions of MLRv, Ander appreciates how controllers focus on "what's happening" with the perceived presence and materiality of *sound* rather than the algorithmic processes of *software*.[64] Put another way, sound becomes a material manifestation of the algorithmic processes embedded within software. In this way, sound can't be understood as separate from the media environment within which it's produced.

Similarly to how the splicing of magnetic tape seemed to allow Schaeffer and others "direct" access to sound itself, the design of button-based grid controllers such as Ableton's "Push" encourages the impression of direct sonic manipulation. In covering the entire surface of the device with buttons—relegating knobs, sliders, and LCD screens to the margins of the interface—the "Push" controller defers the musician's attention to the pads themselves, thus foregrounding the *performance* (process-oriented) aspects of button pushing and finger drumming over the *manipulation* (object-oriented) aspects of sound tweaking and fine-tuning. Redirecting the musicians' focus from the software to the hardware mirrors a shift from the "static" screen to the "liveness" of the performance environment. Producer Decap equates the direct manipulation of the grid with the dynamism of the crowd: "When performing live I literally never need to look at my laptop screen. All of my attention is focused on the *Push*, and the energy of the crowd."[65]

Further, the use of sixty-four buttons (rather than the previous standard of sixteen) allows the grid to structure and constrain the affordances of the software, focusing the musicians' attention on the performative aspects of rhythmic and melodic sequencing. While sequencing drum patterns in real time, for example, "Push" uses color to segment the grid into two separate sections: one containing a sixteen-pad grid to select and demo sounds and another for sequencing these sounds along a rhythmic grid (figure 4.8). This serves as another example of how abstractions of interface elements—whether

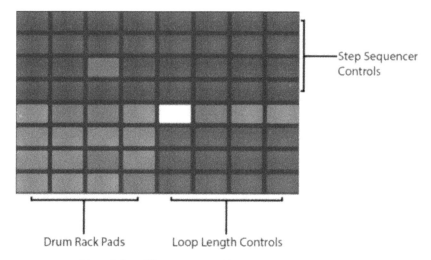

Step Sequencer Controls

Drum Rack Pads   Loop Length Controls

Figure 4.8 Ableton's "Push" version 2 pad layout during drum sequencing. The grid is segmented into four four-by-four sections, each corresponding to specific functions

Source: Ableton, "Using Push 2," Ableton reference manual version 9, accessed August 27, 2020, https://www.ableton.com/en/manual/using-push-2/.

in the form of colored grids in the case of music or shapes and symbols in the case of game controllers—focus the user's attention on specific aspects of the creative experience. Despite the maximal options afforded by "Push," deferring the musicians' focus to a single element of the interface helps consolidate creative direction and reduce the potentially debilitating effects of the "digital maximalist" mindset discussed in chapter 1. ▶

Ultimately, both video game and digital music controllers make tangible the design affordances and constraints of the software being controlled. For gamers, the process of abstracting video game mechanics into the letters and shapes of controllers allows players to embody the rules of games and therefore develop the skills required to succeed in gameplay. For musicians, externalizing the mechanics of music software programs allows performers to convey liveness to their audiences and therefore engage with listeners and technology on a more dynamic level.

Controllerism represents a single solution to a perennial question in digital art: how to physically interact with and manipulate creative affordances embedded in screens. The unending development of hardware for engaging with music software has rightly been criticized as an unsustainable model that runs on the desire for commercial profit—a model paralleled in the games industry. However, the fact that users continue to experiment with controllers, constantly challenging themselves to learn new forms of embodied interaction with their tools, highlights another important value in the experience of contemporary music and games: failure.

The necessity of failure is obvious in the case of gaming, a medium that teaches players to face (virtual) death repeatedly. It's through the unending process of death and resurrection that players learn from their mistakes to develop the skills necessary to "beat the game." Recently, the proliferation of controllers in media production and performance has allowed the built-in possibility of failure and imperfection to bleed into the realm of digital music. DJ Tobias Van Veen sees musical controllerism as a way of enhancing the "human" element of performance: "Without risk of fucking-up, there is no need for the human. . . . [But] controllerism offers possibilities here as a way forward, by which I mean controllerism also entertains virtuosity."[66] Composer Kim Cascone describes failure as "a prominent aesthetic in many of the arts . . . reminding us that our control of technology is an illusion, and revealing digital tools to be only as perfect, precise, and efficient as the humans who build them."[67] Rather than praising the agency and virtuosity of the human over technology, liveness is evidenced instead in the potential for failure inherent in the process of navigating new relationships with technology.

Failure contradicts prevailing ideologies of innovation and progress within the design and technology industries. Each year, Apple releases swaths of computing devices, promising to make the lives of consumers better through user-friendly designs that are easy to navigate and seemingly fail-proof. Likewise, web designers and UX professionals adhere to the "don't make me think" attitude, in which familiar models of interaction are borrowed from existing media to prevent the user from being cognitively or physically challenged in any way.[68] Here "digital" tools stored in the "cloud" are marketed as catch-all solutions to the problems that exist in our material, "analog" world. Friedrich Kittler once said that "there is no software," reflecting on how automated systems are designed to erase the mechanisms through which they work.[69] Yet, as I've discussed throughout this chapter, both the historical relationships between media forms and the practical techniques used to navigate emerging technologies exist in a feedback loop. Through play, musical processes become materialized. It's more fitting, then, to say that there is no software without hardware.

Here we might find a kernel of controllerism's utopian impulse, one that's not so different from the archival work of vinyl hip-hop heads and video gamers. How might improvisation with machines provide a way of imagining technology as a platform for alternative storytelling, thus contradicting common conceptions of technology as agents of control?[70] Or how might controllerism be heard as a form of procedural listening, the performance of technological failure introducing a mode of humanity that doesn't begin with static understandings of objecthood?[71]

Especially following the Me Too movement in the late 2010s, controllerism was embraced as a platform for self-transformation, representation, and social justice. Controllerist legend Laura Escudé's identity as an artist has been defined by these types of transformation, or "transmutation," as she defines it through her Transmute Accelerator live electronic music performance workshop and her 2018 EP, *Transmute*. From her cyborg "Alluxe" persona as the controllerist for Kanye West, to her multi-instrumentalist violin virtuoso work as Laura Escudé, she takes inspiration from artists who "aren't afraid to fail" and regularly describes the excitement that comes with risk-taking and the potential for failure in improvisational electronic music.[72] Twitch icon and "Push" virtuoso dolltr!ck creates video tutorials to bring controllerism to a much broader demographic of musicians, including young women, K–12 students, and economically disenfranchised populations. In addition to her connection to the Twitch streaming platform (which primarily houses video game streams), her website slogan, "enter the dollhouse," analogizes her controllerism performances with the "magic circle" of gameplay, a temporary world within another world, "dedicated to the performance of an act apart," as Dutch historian Johan Huizinga defines the term.[73]

Escudé, dolltr!ck, and the countless other female, nonbinary, and POC controllerists who emerged in the 2010s attest to an increasingly diversifying demographic within the electronic music community and reflect a musical variation of what game expert Bonnie Ruberg calls "queer play," in which failure is embraced for its transformative work, and players change games and their meanings through the playful subversion of rules and mechanics.[74] These controllerists employ what performance theorist Jonathan Bollen calls "queer kinesthesia," reclaiming through movement what's often perceived as the male-dominated space of performance in electronic music and video games "through a marshaling of kinesthetic resources that disarticulate ways of moving from the demand for consistently gendered performance."[75] Following Kiri Miller's ideas about the subversive potential of motion capture interfaces in gaming, their work teaches performers "to regard their own bodies as both interfaces and avatars, a radical change from the established gaming paradigm of using a game controller to direct the actions of an on-screen avatar."[76] In doing so, they present a model for controllerist performance that subverts the fundamental masculinist goal of "control" in music and games.

Exposing the potential for failure at the root of all forms of mediation, controllerism represents an instance of a twenty-first-century digital culture in the process of resisting the perennial narrative of technological process. Like parallel movements in interactive media—net.art, indie video games, glitch aesthetics—controllerism embraces vulnerability as a prevailing ethic of HCI. In each case, the imperfections of the individual operator and the software itself become evidence of liveness. Technological change, in this context, is not simply about developing new, shiny "digital" objects but also playfully experimenting with the embodied, "analog" processes ever present in music and media production. In an era of increased technological control, dominated by proprietary software, global surveillance systems, and the ubiquity of "smart" media, these technologies of play remind us that music, like many of the games we play, consists of rules that are designed to be broken.

5

There's an App for That

Music making is often directly linked to sound production. Conventional
wisdom assumes that music is made when a sounding note leaves the instrument that produced it. Even musicians who depend heavily on technology
tend to think this. Until now, this book has dealt with cultural communities that distinguish themselves *as* musicians precisely by the sounds that
they produce with technology. In the maximalist setups of Flying Lotus and
Rustie, software such as *FL Studio* and *Live* are employed to create EDM and
hip-hop music. In the minimalist setups of *Max* users, custom musical processes are created to experiment more fully with the parameters of sound
production. In controllerist performances, the visceral interactions of live
sound production are valued as fundamental elements of what constitutes
music. While the use of technology among musicians in each case has introduced questions about whether digital music can have the same value as, say,
music made by a classically trained violinist, music making is perceived, in
any case, to occur at the physical or virtual point of sound production.

Contrastingly, music making on devices such as smartphones foregrounds
not the actual sounds produced by the instruments but rather the physical
actions and seemingly invisible design affordances that guide the mundane
ways in which we interact with these devices on an everyday basis. In contrast to controllerists who use hardware controllers to distinguish themselves
from commonplace users of technology, mobile media users celebrate the
disintegrating distinction between expert "producers" and nonexpert "users"
facilitated by devices such as the iPhone. The line between creative production and media consumption is further blurred by microtransactional
affordances such as the *App Store*, an online shop where users can purchase
new apps and add-on content for existing apps on the iPhone. Since the
case of the iPhone illustrates that music is not perceived to occur simply at
the point of sound production, we're left with two key questions: what's the
nature of mobile music making, and how can we understand the nature of
music making more broadly when sound production (and the device user's
expertise at it) is not perceived to be the core element of music?

Push. Mike D'Errico, Oxford University Press. © Oxford University Press 2022.
DOI: 10.1093/oso/9780190943301.003.0006

In both its corporate marketing and public perception, the Apple iPhone is a device whose user-friendly design and seemingly intuitive touchscreen mechanics equalize the skill levels needed for everyday productivity tasks and creative music production. I suggest that music making in the context of mobile media involves acts of both production (through the physical gestures and design affordances of music-making apps themselves) and consumption (through the generative structure of social-media sharing, as well as add-on content that can be purchased separately from the app). Specifically, I analyze the intuitive iPhone and app design affordances that allow the adaptation of the device in mundane, everyday interactions. These affordances include direct manipulation, touchscreen control gestures, elementary physics in app design, and multitasking capabilities. By analyzing how mobile media software and app design facilitates nonexpert music production practices, it's possible to understand digital music as made up not only of *material* technologies (instruments, controllers) and performance spaces (dance clubs, concert halls) but also of process-oriented *experiences* that align with consumption practices inherent in capitalism in the early twenty-first century.[1] Before addressing the design affordances themselves, I outline the paradigm shift in digital media production from an emphasis on specialized skill to a form of creative production that also embraces intuitive, nonexpert knowledge. This shift can be detailed in twentieth-century developments in music technology, as well as the philosophy of personal computing in the 1980s—the design and marketing doctrine that eventually gave rise to the invention of the iPhone.

With the rise of sound reproduction technologies throughout the twentieth century, music making became increasingly available and accessible to a broader demographic of nonexperts.[2] Or at least that's what the music technology companies tell us. Ethnomusicologist Timothy Taylor discusses how advertisers promoted the player piano as a technology that fostered a democratization of both ability (in that it required no specialized skills to operate) and availability (in that it was more affordable than a grand piano).[3] By convincing consumers that the player piano allowed them a level of "freedom from technique," player piano advertisers marked "the beginning of the transformation of the musical experience into an object of consumption."[4] Mark Katz considers how the phonograph similarly facilitated "amateur" music making during this time, while Kiri Miller has examined production practices in the context of "amateur-to-amateur" internet platforms such as YouTube.[5] In each case, communities of nonexperts formed as a result of increased financial accessibility and usability in design.

Similarly, in the history of twentieth-century computing, there was a gradual shift from a mode of thought that we might call specialized computing, in which technical skill is valued among particular communities of experts, to a newer paradigm that became known as personal computing. Specialized computing emerged in the midst of the technological competition occurring between nations during and immediately following World War II.[6] The world's first electronic computers were expensive, required specialized technical knowledge to operate on a basic level, and often could only be found in specialized venues such as university research labs.[7] For example, the first commercially available electronic computer, the Ferranti Mark 1 (1951), sold only ten units in total, for £83,000 each, or almost $3 million US in today's terms (figure 5.1).

Personal computing emerged following the development of the microprocessor in the early 1970s. In response to the specialized nature of computers after World War II, the underlying premise of the personal computing philosophy is the belief that the affordances of technology should

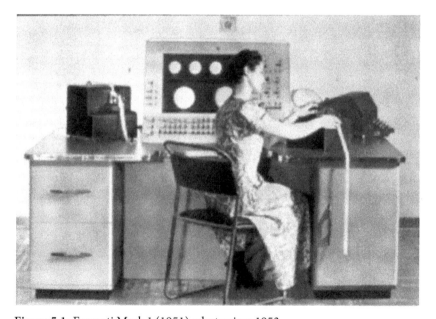

Figure 5.1 Ferranti Mark 1 (1951), photo circa 1953

Source: Computer History Museum, September 8, 2005, accessed August 31, 2020, https://www.computerhistory.org/chess/stl-430b9bbe6b611/.

be commercially available at an affordable price and usable by the broadest demographic possible.[8] In contrast to computers such as the Ferranti Mark 1, companies such as Xerox and IBM aimed to create smaller, more affordable computers that could be taken outside the computer lab and into users' everyday lives (figure 5.2).

Apple contributed to the personal computer revolution on a mass commercial scale in the late 1970s. During this time, Apple co-founder and

Figure 5.2 Xerox Alto I CPU with monitor, mouse, keyboard, and five-key chording keyset (1973)

Source: Computer History Museum, n.d., accessed August 31, 2020, http://www.computerhistory. org/revolution/input-output/14/347.

designer Steve Wozniak regularly attended meetings of the Homebrew Computer Club, an informal group of Silicon Valley computer enthusiasts. Wozniak was so inspired by the group that he began designing what would become Apple's first personal computer after just one meeting.[9] Instead of designing complex interfaces only accessible to the expert computer user, Apple asked, what are humans *naturally* good at, and how can we embed those affordances in software and hardware? To appeal to the broadest demographic of computer users, the company established a design philosophy centered on three core tenets: the marketing of "everyday" creativity, visual aesthetics over computational power, and minimalist design.

By employing a marketing strategy that focuses on the everyday creativity afforded by its products, Apple continues to represent the apex of user-friendly populism in computing. As the slogan from the original Macintosh (1984) commercials attests, Apple's design philosophy has been eternally invested in the democratizing goal of creating a "computer for the rest of us" (figure 5.3).[10] This goal has inspired nonexpert artists and musicians to embrace the Mac operating system in their creative work, as an easier-to-use counterpart to Microsoft's Windows. With the introduction of *iMovie* video editing software in 1999, Apple products were promoted to be "the next big

Figure 5.3 Advertisement for Apple Macintosh (1984)

Source: prattcomd520, "User Centered Design," accessed April 30, 2016, http://dosislas.org/pratt/spring16/comd520/user-centered-design/.

thing" in personal computing, turning their users into "both the director and the producer"[11]—a marketing rhetoric that promoted what digital media researcher Jean Burgess calls "vernacular" creativity.[12] According to Burgess, the technical design of the original Macintosh and the *iLife* suite, as well as the advertising campaign surrounding them, "constructed its users as effortlessly creative rather than extraordinarily so."[13] In contrast to specialized technical skill, vernacular creativity encouraged a playful remixing of content that merged amateur creativity, technology, and everyday experience.

In addition to promoting the democratizing power of vernacular creativity in its products, Apple also redirected the focus of computing from computational power and technical skill to visual aesthetics. The original iMac, introduced in 1999, embodied this shift in its focus on the color and aesthetics of the device above and beyond what Steve Jobs called the technical "mumbo jumbo" of the hardware.[14] Available in five different colors, these computers were promoted as creative platforms for users to express themselves, closely aligning the company's marketing and design schemes (figure 5.4). Since this moment, for many users, the intimate processes of interaction have become more important than the goal-oriented tasks of computing. These users respond to Apple's goal of fostering increasingly intimate relationships between humans and their computers, reflecting the gendered nature of computing more generally.[15] This message is apparent in many of Jobs's product launch presentations, as everything from the iMac to the iPod is described as "beautiful," "gorgeous," and even "delicious."[16]

The emphasis on visual aesthetics is most noticeable in Apple's signature minimalist design aesthetic, summarized by the Dieter Rams–inspired philosophy of Apple's chief design officer Sir Jonathan Ive in the phrase "good design is as little design as possible."[17] The 2015 MacBook laptop, for example, includes a streamlined, rounded rectangle casing that provides the guiding design framework for everything contained within it, from the keyboard enclosure and mouse trackpad to the rounded square software icons and screen notifications. Interface elements on the hardware consist of three simple elements: a single peripheral output on the side of the machine, a flat trackpad, and a black-and-white rectangular keyboard (figure 5.5). Cut from a single piece of aluminum, the unibody exterior of the device accentuates a soft, clean surface, enhancing the minimalist control surfaces present on the computer. In reducing the number of *hardware* elements on the device, the MacBook focuses the users' attention on the *software* processes taking place on-screen.

Figure 5.4 Apple's "Yum" ad campaign (1999)

Source: Dylan Tweny, "Vintage Posters Highlight a Century of Innovation," *Wired*, February 3, 2011, accessed August 31, 2020, https://www.wired.com/2011/02/vintage-posters/.

Apple's minimalist design aesthetic has thrived with the emergence of mobile media. The history of personal computing—as with many technological innovations of the twentieth and twenty-first centuries—is a story of increasing miniaturization, portability, and accessibility. As devices get smaller, the body learns to confine its physical gestures so as not to exceed the confines of the screen. Perhaps more than any other device, the iPhone foregrounds the aesthetics of HCI by solidifying Apple's minimalist design philosophy. This is most noticeable in the control mechanisms of the device. While most smartphones at the time of the iPhone's release in 2007 contained button-based keyboards that took up half the phone's front panel space, Apple removed the physical keyboard entirely in favor of direct touchscreen control (figure 5.6). According to Ive, this was Apple's primary design strategy in constructing the iPhone, as it would encourage users to

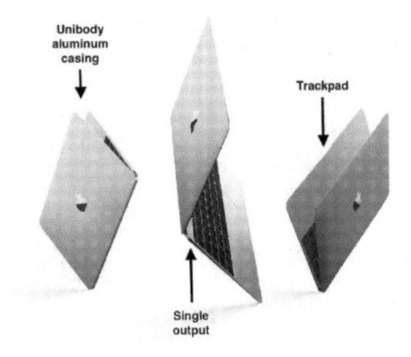

Unibody aluminum casing

Trackpad

Single output

Figure 5.5 Apple MacBook (2015)

Source: Apple, "Macbook," n.d., accessed April 30, 2016, http://www.apple.com/macbook/; annotations added.

defer all mental and physical attention to the tactile space of the touchscreen, rather than the limited number of buttons and switches outside of it.[18] In focusing the user's attention on the visceral connection between the human finger and the smooth glass screen, Apple further promotes vernacular creativity, evoking a sense of direct, unmediated connection between the device and its user.

Together, developments in music technology and personal computing marked a cultural shift in the perceived role of technology among everyday users. Whereas traditional forms of music making and computing valued the development of specialized, technical skills, the new paradigms of non-expert music making and personal computing valued the proliferation of non-specialized creative production. In doing so, these developments continue to encourage shifting understandings of the role of music in everyday life. Mobile music production with iPhone apps serves as one of the most

Figure 5.6 Apple iPhone (2007)

Source:"Retro Phone Review: the original Apple iPhone," Tech Radar, July 20, 2015, accessed August 31, 2020, https://www.techradar.com/news/phone-and-communications/mobile-phones/retro-phone-review-the-original-apple-iphone-1299387.

important developments in music software, technology, and personal computing in the early twenty-first century. In this context, we might ask, how do the design affordances of musical apps affect the behavioral processes of everyday music making and computing? What happens to conceptions of music if entire songs can be composed and distributed with nothing but a mobile phone and a one-dollar piece of software? Has the so-called democratization of technology made music available to everybody, or has it diluted music's claim as a specialized cultural practice?

The advent of mobile media represents the apex of personal computing values, sparking a renewed interest among corporations and DIY musicians alike in the democratization of technical ability and affordability. The iPhone embeds the tacit and elementary bodily comportments of computer users into the physical design of the device itself, as well as the software apps stored on it. Let's consider four of the most common design affordances in mobile media devices: simple touchscreen control gestures, elementary virtual physics in apps, the impression of direct manipulation, and multitasking capabilities. These affordances exploit forms of physical and cognitive interaction that humans are naturally familiar with and, as such, require no specialized training or technical skills. The iPhone thereby blurs the boundary between creative production and the mundane tasks of everyday life. As the personal computer literally and metaphorically moves from the lab to the home to the pocket, the hardware seems to gradually disappear, the iPhone ingraining itself into the everyday interactions of its user.

The employment of simple touchscreen control gestures is a unique design affordance introduced by the iPhone. Most notably, the touchscreen aligns creative production with the naive gestures of the human body, rather than the specialized technique of a musical instrument. Apple has gone so far as to copyright specific physical actions, thus standardizing touchscreen gestures such as tapping, dragging, flicking, swiping, double-tapping, pinching, touching and holding, and shaking. In order to standardize these gestures, Apple regularly publishes a set of "iOS Human Interface Guidelines" that instructs designers in how to employ specific actions for specific purposes.[19] For example, "flick" is to be used for scrolling quickly through a set of content, "shake" will undo or redo an action, and "pinch" is used to zoom in or out. Designers are specifically discouraged from introducing new gestures into the app unless it's a video game (figure 5.7). As Apple claims, "In games and other immersive apps, custom gestures can be a fun part of the experience. But in apps that help people do things that are important to them, it's best to use standard gestures because people don't have to make an effort to discover them or remember them." These guidelines highlight the proprietary nature of iOS as a dominant operating system, as well as a radical departure from traditional "musical" techniques, skills, and forms of interaction.

Music app developers have been quick to embrace touchscreen gestures to facilitate more intuitive music making with mobile devices. Native Instruments' *Traktor DJ* promises the ability to "touch the groove" by using tapping and pinching to edit directly audio waveforms directly (figure 5.8). More abstract apps such as *Patatap* fragment the iPhone screen into twenty-four blocks that trigger audio samples when tapped (figure 5.9). Reason Studios' *Figure* uses a trackpad-style control mechanism that allows musicians to create rhythmic variety in their musical patterns by swiping along an x–y grid (figure 5.10). Others employ touchscreen mechanics for pedagogical purposes. The *Clapping Music* app, for example, challenges the smartphone user to tap the screen in time with Steve Reich's infamous minimalist composition "Clapping Music" (figure 5.11).[20] In order for users to develop the digital (and musical) literacies required of apps, they must develop

Figure 5.7 Apple iOS Standard Gestures

Source: Apple, "Gestures," in *Human Interface Guidelines*, accessed September 21, 2021, https://developer.apple.com/design/human-interface-guidelines/ios/user-interaction/gestures/.

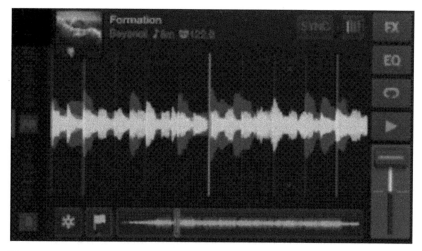

Figure 5.8 Native Instruments, *Traktor DJ* (2013) (screenshot taken by author).

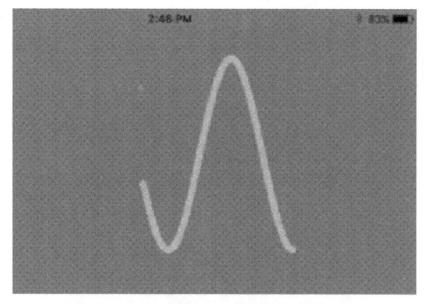

Figure 5.9 Jonathan Brandel, *Patatap* (2014) (screenshot taken by author).

a habituation of response through repeated practice. Like many forms of interaction with mobile media, music production apps thus rely on the omnipresence of the iPhone in the everyday routines of the user.

In order to teach users how to incorporate these new control gestures into their physical comportments with mobile media, app designers incorporate elementary physics—programming scripts that cause *virtual* interface elements to simulate the behavior of *physical* objects—into app UIs. This design affordance gives users the impression that they interact with physical objects, rather than a flat glass screen. In order to be effectively operated, post-WIMP interfaces—that is, interfaces that diverge from the "window, icon, menu, pointer" layout from desktop computing—require users to internalize viscerally the affordances of the app in the absence of more traditional computer interfaces such as the mouse and the keyboard.[21] This internalization is achieved through the elementary physics imbued by app design, which includes sensations such as inertia and springiness among interface elements. For example, a user encounters elementary physics in action when the color of an app icon darkens once it's pressed. This effect provides the illusion that the virtual objects present throughout the device (buttons, in this case) have mass.

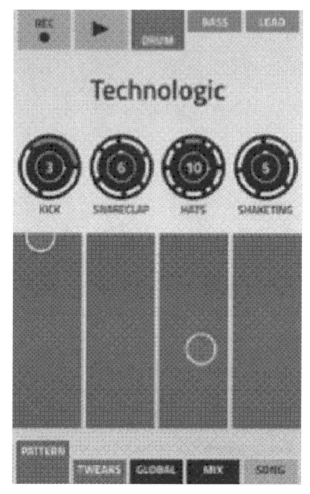

Figure 5.10 Reason Studios, *Figure* (2012) (screenshot taken by author).

These "naive physics" can also extend to our embodied memory of hardware such as computer keyboards that are simulated in the iPhone GUI.[22] This means that users learn to navigate the device by routinely practicing new sets of physical gestures and by translating to the smartphone their embodied knowledge developed from interactions with past media forms— pressing buttons on an analog telephone, for example. The skeuomorphic metaphors and visual analogies embedded within apps afford users the illusion that the apps are based on intuitive forms of interaction, rather than complex software code or musical technique. Thus, users are encouraged to

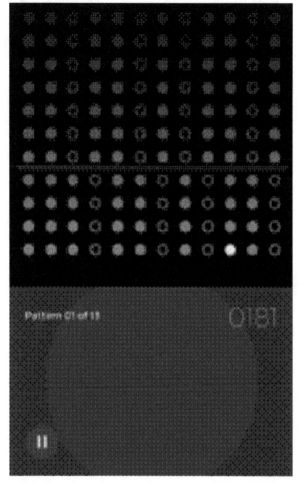

Figure 5.11 *Clapping Music* (2014) (screenshot taken by author).

embody further the physical gestures of the iPhone touchscreen in their everyday lives.

Abstracting *physical* gestures into *virtual* touchscreen gestures through elementary physics is crucial for music production apps, which often succeed or fail based on how well they emulate the tactile affordances of physical instruments. Native Instruments' *iMaschine* is a good example of this, designed to emulate the classic drum machines and samplers used by hip-hop and EDM producers. The focus of the interface is a sixteen-pad grid of square buttons used to control sound samples (figure 5.12). As in samplers

Figure 5.12 Sample pads in *iMaschine* (screenshot taken by author).

such as the Akai MPC, each pad is assigned an individual sample that can be edited by adjusting length and volume or adding effects. Extra features of the app, including the ability to construct rhythmic patterns, arrange songs, and set playback parameters, are relegated to the outer margins of the screen. Like the design of the iPhone itself, this layout focuses the users' attention on the *process* of "pushing" the buttons on the grid rather than the *objects* simulated. ▶

In order to model previous techniques of beat production, for example, elementary physics in the app are directly mapped to physical gestures required by the analog machines being simulated. When the virtual pads on *iMaschine* are "pushed," they illuminate and display a 3D gradient around their edges, giving the impression of visual depth and weight to GUI elements. Similarly, moving a virtual fader by sliding a finger across the glass screen of the iPhone gesturally emulates the smoothness of adjusting a fader on an analog mixer. These synesthetic illusions allow the user to internalize the abstract metaphors and elementary physics on which the app's core mechanics are based.

Together, touchscreen gestures and elementary app physics encourage the development of another common iPhone design affordance: direct manipulation, which relies on the foregrounding of material content over extraneous administrative interface elements.[23] The iPhone gives the user the impression of immediate, direct manipulation of content by combining the physicality of touchscreen gestures with maximal screen real estate made up of images, audio waveforms, and video clips, rather than GUI elements. Despite the small-screen hardware, increased screen resolution facilitates even greater information transmission and detail of content, allowing the user to interact with the maximum number of interface elements. This feature allows iPhone users to edit images with a degree of clarity and precision similar to using a desktop software program such as Adobe *Photoshop*. Further, the operating system tends to distribute all content on the same horizontal surface, cementing the idea that content creation and editing are continuous rather than interrupted. As the typical navigational elements such as "Back" buttons, text input bars, and arrows are minimized, the content becomes the interface, and users feel that they're naturally exploring the device rather than being led down a goal-oriented path (figure 5.13).

By removing extraneous interface elements from the screen, iPhone apps also foreground the process-oriented mechanics at work behind the interface, allowing the user to play with the rules of the software in various ways. This form of playful interaction, like the "locomotor play" of video gaming and music making described in chapter 4, enhances the feeling of direct manipulation in a few ways. First, audio sampling apps expose the waveform of musical tracks, giving the musician the ability to stretch, loop, or cut up audio by tapping or swiping the screen. In the "Edit Sample" screen of *iMaschine*, for example, users can set the start and end points of a sample by sliding a finger across the screen, rather than using the less intuitive WIMP interface to zoom in on a waveform and drag the pointer across the sample (figure 5.14).

Figure 5.13 iOS *Photos* app (screenshot taken by author).

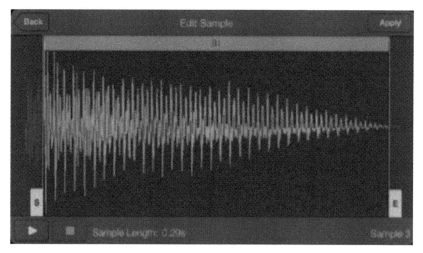

Figure 5.14 *iMaschine* "Edit Sample" screen (screenshot taken by author).

Second, music apps often abstract more traditional musical instruments into shapes, providing game-like experiences. Apps such as *Musyc* (2013) and *Scape* (2012) apply sets of rules to simple shapes, resulting in complex rhythmic and harmonic relationships as the shapes interact with each other (figure 5.15). Combining the sampling capabilities of digital tools with the playful systems of video games, iPhone apps provide users with a feeling that they're interacting naturally with a process-oriented experience, rather than

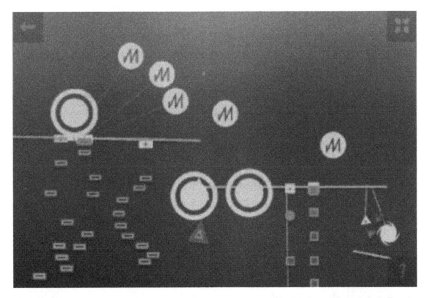

Figure 5.15 Fingerlab's *Musyc* iPhone app (2013) (screenshot taken by author).

the clunky tool in their hands.[24] Whereas the GUI originated in the realm of labor, with its focus on function, interface researchers Søren Bro Pold and Christian Ulrik Andersen note that "the new app interface clearly has its roots in (digital) culture with an *aesthetic* interface, inspired by games, software culture and cultural interfaces in general."[25] In contrast to the critiques that usability in design "dumbs down" the user experience, direct manipulation serves as a condition of possibility for intuitive forms of nonspecialist creativity in music production.

However, direct manipulation also presents a paradox for app designers. Just as the touchscreen presents new possibilities for gestural manipulation, the users' focus on the small screen anchors the body by fixing attention on the software and preventing a more fine-tuned degree of physical control with the device. After the introduction of smartphones in the late 2000s, the "iPhone zombie" became commonplace as large crowds of people were observed having all eyes glued to smartphone screens as fingers rapidly typed away.[26] For controllerists, this situation represents the worst consequence of usability in personal computing, as the usability of the iPhone interface reduces mobile music producers to button pushers who are no different from average mobile media users checking their email or texting their friends. In contrast, iPhone apps embrace the disintegrating distinction between creative and mundane work on the device. This brings me to the final major design affordance of iPhone apps: creative multitasking.

Since touchscreen gestures, elementary physics, and direct manipulation affordances embedded within app design are shared across the various activities with which the iPhone user is engaged, multitasking skills are crucial to everyday mobile media usage. The act of composing music on smartphones, for example, is often done in conjunction with a range of other activities, from checking email to scheduling appointments and browsing news feeds. As a result, app designers must consider what media theorist Robert Rosenberger calls the "field composition" of both the musical experience and the material context of the musicians' everyday life.[27] That is, the designer must consider how the affordances being offered to users can structure their general field of awareness in everyday life.

On one hand, the smartphone encourages a heightened focus on the device itself, allowing users to ignore their external environment. It's no coincidence, then, that many of the promotional materials for music production apps portray the musician interacting with the software in public venues typically perceived to be difficult to navigate socially. The promotional video for *Figure* depicts a man entering a crowded subway train, eyes fixed on his iPhone and over-the-ear headphones safely sheltering him from the noise surrounding him (figure 5.16).[28] As the video time elapses through the

Figure 5.16 *Figure* promotional video (2012)
Source: "Figure for iPhone and iPad," Reason Studios, uploaded April 4, 2012, accessed August 31, 2020, https://www.youtube.com/watch?v=gLLjRH6GJec.

duration of his commute, he never looks away from the app. The ad presents *Figure* as both a distraction from the outside world and a tool for increasing one's focus on the creative process. It's the quintessential image of the always-online individual simultaneously completely enmeshed in the social network and absolutely separate from it—a condition that's come to define mobile media usage.[29]

On the other hand, the UX workflows of apps are designed to allow for multitasking in response to the demands of life outside of the smartphone. What I call mechanics of interruptibility allow the iPhone to become embedded into users' everyday lives, precisely by encouraging usage in small doses. Initially coined by game theorist Jesper Juul to describe how play becomes intertwined with everyday routines, interruptibility is employed as a mobile design strategy to maximize user productivity while working with multiple apps simultaneously.[30] The larger screens of laptops facilitate multitasking by segmenting the screen into multiple "windows." Swiping gestures on the trackpad allow the user to switch between active software programs, reorganize windows currently visible on-screen, and divide the screen into multiple "desktop" workspaces (figure 5.17).[31] ▶

The iPhone achieves this multitasking workspace by organizing information spatially in the manner of folders, maximizing the real estate of the

Figure 5.17 Apple Mac OS X "Show Desktop" trackpad gesture reveals active software programs (screenshot taken by author).

smaller screen by adding multiple layers of depth to the interface.[32] When multiple tasks are performed, the interface collapses into a system of layers, allowing the user to switch between apps with ease and speed. At any moment, a double tap on the home button reveals an accordion-like menu of app "tabs" at the top of the screen that serve as navigational "breadcrumbs" (figure 5.18). In this context, composing and producing music occur within the device and the interface and are therefore given no more or less value than texting, checking email, or playing a video game.

By allowing the user to "pause" the music-making process in this way, the iPhone affords a continuous production and uninterrupted creative

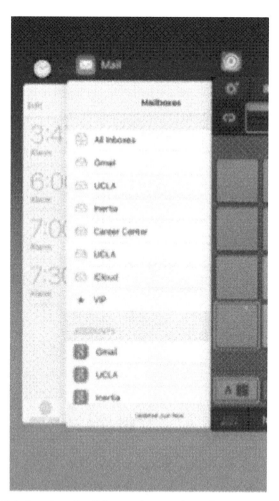

Figure 5.18 iOS "App Switcher" feature (screenshot taken by author).

flow among the various tasks with which the mobile media user is engaged. Imagine you're making a beat in *iMaschine*. You receive an important email, and a banner notification appears at the top of the screen. As the music plays, you tap the notification and are immediately taken out of *iMaschine* and into the *Mail* app. At the same time, a red banner affixes itself at the top of the screen, letting you know that an active recording session is taking place in *iMaschine* and that you're able to return to the app simply by tapping the red banner. Here, despite the somewhat complicated multitasking going on, the use of banners at the periphery of the user's attention serves as a cognitive glue for the user, creating the feeling of uninterrupted, perpetual production. Text message banner notifications are even less disruptive, as the user can respond to the text in the notification itself, without having to leave the current app. After receiving the text message, the user swipes down on the banner, and a text box appears. Then *iMaschine* darkens and blurs, momentarily foregrounding the act of texting over music production (figure 5.19). In these examples, mechanics of interruptibility illustrate the convergent nature of production on the iPhone, heightening the musician's feeling that the creative process of music making is as ubiquitous and mundane as checking email.

The design affordances previously outlined—simple touchscreen control gestures, elementary virtual physics, the impression of direct manipulation, and multitasking capabilities—increase the appeal of the iPhone for a broad demographic of nonexpert users by foregrounding intuitive forms of interaction over specialized technical skill. In doing so, the previously isolated practices of creative music making and everyday production tasks are integrated within the immediately accessible and all-encompassing smartphone. Of course, even before the rise of the iPhone, the design of personal computers was already moving in the direction of increased mobility and pervasive use. In his 1991 article "The Computer for the 21st Century," computer scientist Mark Weiser foreshadows an emerging age of "ubiquitous computing" in which computers would "weave themselves into the fabric of everyday life until they are indistinguishable from it," thus fading to the background of users' attention.[33] According to Weiser, rather than being a *technological* issue, the desired "invisibility" of the computer is a *psychological* issue related to the development of tacit knowledge. As he states, "whenever people learn something sufficiently well, they cease to be aware of it."[34] The now standard physical actions of swiping and tapping the iPhone screen highlight the tacit corporeal knowledge required of media consumption and

Figure 5.19 Multitasking and interruptibility on the iPhone (screenshot taken by author).

production in the age of mobile media, constituting the core characteristics of the everyday, vernacular creativity that has come to define personal computing.

Mobile app developers have also embraced marketing rhetoric that emphasizes ubiquitous computing as a democratizing force. iPhone music app developers Robert Hamilton, Jeffrey Smith, and Ge Wang consider the rise of "ubiquitous music" as a consequence of mobile media usage, as apps "increasingly transport us into a world where we do not have to immerse ourselves in computers, but instead take computing into our physical world and

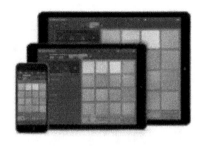

Figure 5.20 *iMaschine 2* promotional material

Source: Native Instruments, "iMaschine 2," accessed August 28, 2020, https://www.native-instrume nts.com/en/products/maschine/maschine-for-ios/imaschine-2/.

Figure 5.21 *GarageBand* iOS promotional material

Source: Apple, "iOSGarageband," accessed August 28, 2020, http://www.apple.com/ios/garageband/.

nearly every part of our daily life."[35] Native Instruments' *iMaschine* adver-tisement offers musicians the ability to "make music anywhere," embedding full-fledged audio sampling and editing capabilities in a five-dollar iPhone app (figure 5.20). In reference to their *GarageBand* iOS app, Apple expresses to consumers that "The world is your stage. This is your instrument" (figure 5.21).[36] Reason Studios' *Figure* is described as a "fun music-making app for instant inspiration," allowing the musician to "create music in no time with *Figure*'s dead-easy touch interface" (figure 5.22).[37] While previous visions for the future of personal computing, such as VR, involved immersing individ-uals *inside* a computer, ubiquitous computing "brings the computer outside,

Figure 5.22 *Figure* promotional material
Source: Reason Studios, "Mobile Apps," accessed August 28, 2020, https://www.reasonstudios.com/mobile-apps.

into our daily, lived experiences," as media theorists Adriana de Souza e Silva and Daniel Sutko note.[38] In contrast to software such as *Pro Tools*, which emphasizes the professional and specialized nature of the program, music app marketing highlights how the rhetoric of ubiquitous computing in the lineage of personal computing has shaped consumer demand for the development of nonexpert production tools.

This situation brings up some questions. When mobile app developers build apps, and non-"musicians" make music with apps, is this really music making at all? Or are consumers simply buying into a specific brand experience? Is the question of what constitutes "real music" even useful anymore? I suggest that the intuitive, natural, and ubiquitous affordances embedded within apps align the practices of mobile music *production* and digital media *consumption*. This is evident in the context of juggernaut social media apps such as TikTok that dominated the charts in the late 2010s and whose content is derived from user-generated "remixes" of various sorts. However, apps specifically dedicated to "Music" making (with a capital M) function through similar design mechanics. In doing so, it's possible to understand music increasingly as deeply intertwined with the maximalist consumption practices described in chapter 1.

Mobile media exemplifies a dominant understanding of technology in the context of consumer capitalism, the idea that consuming technology can solve our problems and provide instant gratification. When Apple first introduced the iPhone, the issue Jobs saw in existing phones was that they all included a bunch of plastic buttons that remained the same for every application. By removing the physical keyboard and increasing the screen size of the device, the iPhone GUI could literally become anything, depending on the nature of the app. The icon-grid interface layout that's come to define the iPhone reflects this tool-based conception of the device (figure 5.23). It's a Swiss army knife of solutionism, offering to better your health, relationships, productivity, work, and everything else in your life with a single tap.[39]

Similar to the maximalist attitude toward technology described in chapter 1, some view the iPhone as a "tethered appliance" on which users have developed a dangerous overdependence.[40] Others view the iPhone as another useless gadget that reflects the problematic dynamic

Figure 5.23 iPhone iOS 13 (2019) home screen (screenshot taken by author).

of "technological solutionism," in which software perpetuates the very problems that it promises to alleviate.[41] Media theorist Svitlana Matviyenko describes how "the 'needs' come with apps as part of the package, which means the 'solutions' are being sold to us along with the 'problems' they're meant to resolve."[42] Specifically, Matviyenko pinpoints associations between technology and happiness in advertising rhetoric, including Nokia's app store slogan, "Think appy thoughts," or Apple's popular motto, "There's an app for that."[43] In both cases, apps are marketed as catch-all solutions to problems that often arise, paradoxically, from the ubiquitous presence of technology in all aspects of our lives.

In the context of music and media production apps, solutionism manifests in the rhetoric of democratization and accessibility introduced at the beginning of this chapter—the idea that, with an iPhone, anybody can make anything anywhere. Apple claims that the *GarageBand* app, for example, allows users to "create incredible beats no matter where you are . . . with just a few taps." Similarly, with Apple's *iMovie* app, all it takes is "a few taps, a few swipes, and you're ready for your big premiere."[44] The marketing of the *Pages* app is perhaps the bluntest example, with the claim that the software allows you to "effortlessly create stunning documents. . . . Writing has never been easier. Period."[45] By aligning simple touchscreen gestures with professional media production techniques, Apple both perpetuates its marketing ethos of vernacular creativity and appeals to the perennial anxiety that many users hold toward creativity with the personal computer. The seemingly immediate nature of interaction with the iPhone is posited as a catch-all solution to a long-standing unease and awkwardness that many feel when producing art and music with the standard WIMP interface. The title "smartphone" itself blurs the line between expert and nonexpert production, encouraging users that it's not only the phone but also its user that's "smart."

The convergence of creative practices in mobile media production has resulted in the formation of what I call an upgrade culture, in which users simultaneously consume and produce content as part of an endlessly iterative digital economy. This framework for understanding the self-perpetuating organization of mobile media economies was foreshadowed by philosopher Henri Lefebvre as far back as 1967, when he introduced the concept of "controlled consumption." According to Lefebvre, everyday life in advanced industrialized societies has become a "voluntary programmed self-regulation" contained within a "closed circuit" of "production-consumption-production."[46] Translated to the twenty-first-century context of the iPhone, the act of consuming apps perpetuates the very needs the software promises to solve. Lefebvre outlines four principles that characterize societies of controlled consumption: (1) a cybernetic industrial infrastructure integrating and handling production, distribution, exchange, and consumption is developed around the product; (2) the consumption is controlled through programming that closely monitors consumer behavior and the effects of marketing through tracking and surveillance; (3) controlled obsolescence is programmed into the product, limiting its functionality and its durability; and (4) the overall effect of controlled consumption is a significant reorganization and troubling of specific practices of everyday life.[47] These four principles are most clearly demonstrated by the mechanism that drives both production and consumption on the iPhone: the *App Store*.

Apple's *App Store* represents a centralized marketing app, a meta-app that structures the production and distribution of software on iOS devices (figure 5.24). Represented in early iterations by a simple pencil, paintbrush, and ruler icon, the store promises "productivity" above all else, offering almost 2 million apps ranging from self-help tips to video games, word processors, DAWs, and design tutorials. App prices are low, for the most part, anywhere from completely free to ten dollars. In addition to this initial fee, some apps contain "in-app purchases" that either provide users with extended functionality or hook them into long-term use through a subscription model. Significantly, most software developed for the iPhone is created by third-party companies. Therefore, Apple both strictly controls the content and distribution of apps in the store and relies on outside labor to maintain its mechanics. ⊛

Controlled consumption reflects the dominant ideology of consumer capitalism following the rise of software—an ideology "designed to sell not only a particular commodity but consumption itself," according to Taylor.[48] Controlled consumption in the *App Store* succeeds through the deployment of "microtransactions" such as in-app purchases, software subscriptions, and DLC. Recently, Google claimed that the best method for hooking users into the extended life cycle of a product is to fracture the consumer journey into "hundreds of real-time, intent-driven micro-moments," allowing the smartphone user "to act on any impulse at any time" to achieve "immediate results."[49] Through microtransactions in the *App Store*, users' labor is more substantial than the apps themselves in preserving the controlled consumption model.

Figure 5.24 iPhone *App Store* (2020) (screenshot taken by author).

While the apps are either given away or sold for a couple of dollars, in-app purchases exist as either monthly subscription payments or one-time fees of more than a hundred dollars. In iOS games, for example, the rise of the "freemium" model—in which DLC can be purchased by gamers seeking extra levels or other perks—has managed to hook even the most casual players into the extended life cycle of the product. The combat strategy game *Clash of Clans* grosses more than $1 million a day, largely from DLC purchases that allow players to strengthen their village.[50] In music apps, DLC often comes in the form of sample packs or expansion plug-ins for the software. The *iMaschine* "Expansions Store," for example, regularly releases new virtual drum kits, synthesizers, and samples with which the musician can build tracks (figure 5.25). In each case, in-app purchases extend the act of acquiring software from a single event to a gradual process. Perpetual *consumption* becomes a primary mechanic of both gaming and music *production*.

Of course, the iPhone is more than just a tool of controlled consumption, simply providing users with instant fixes to their personal problems. It's also a generative creative platform.[51] In addition to DLC sold by developers following an app's release, the ability of users to develop and share original content is often built into the app's design. Designers of music apps are especially

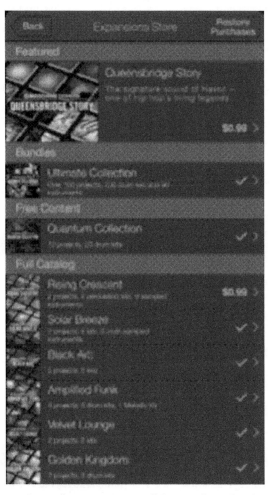

Figure 5.25 *iMaschine* 2 "Expansions Store" (screenshot taken by author).

aware of the benefits in employing "user-generated content," which typically involves musicians sharing their creations with other users and, in turn, promoting the app through social media. Akai's *iMPC Pro* allows the user to upload tracks created in the app directly to social-media networks such as SoundCloud (figure 5.26). For a few years, *Figure* integrated the *Allihoopa* social music platform (now integrated into Reason Studios), allowing users to share their creations with thousands of other "campfire crooners," "table top drummers," and "shower DJs" involved in mobile music creation. Like the rhetoric surrounding *Ocarina*, the *Allihoopa* platform claimed to "bring all of us, everyone, together for a free and open exchange of music pieces, ideas and people, all united in 'doing music' just for the sake of it."[52] If in-app purchases epitomize the controlled consumption model inherent to the *App Store*, affordances for creating user-generated content highlight the generative potential of the iPhone as a creative platform.[53]

As the logics of gaming increasingly infiltrate various media—from cinema to video games and mobile media—the aesthetics of media production cultures converge. In contrast to computer scientist Jonathan Zittrain's notion of the iPhone as *simply* a tethered appliance, the convergent nature of apps within the smartphone also constitutes what media theorist William Merrin calls a "hyperludic" interface: a digital gadget designed for endless functioning and with which we can never exhaust our play.[54] In this context,

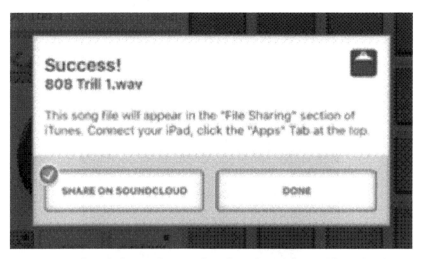

Figure 5.26 Akai Professional, *iMPC for iPhone* (2013) (screenshot taken by author).

the intuitive physical gestures and elementary design affordances of music production apps play a crucial role in facilitating the instantly networked, informal sharing economies of the "remix culture" that has come to define creative production on digital platforms.[55] Think about how "liking," "retweeting," "sharing," and "favoriting" social-media posts, for example, perpetuate the viral nature of online content distribution. Here the ludic affordances of the interface continuously generate more forms of playful interaction. Music making, in this context, is more than the simple acquisition of goods "but the mean mode of relating to goods, and to one another," as Taylor suggests.[56] These examples embrace how mobile media production can empower the user and encourage a sense of social togetherness, rather than creating a sea of mindless consumers.

The popular *Ocarina* app (2008) is a case in point. Just a year after Apple released the iPhone, mobile app development company Smule launched *Ocarina*, the first app to turn the smartphone into a musical instrument. The app offers the best of both worlds for music production and communication. Upon launching *Ocarina*, the user can view a tutorial on how to make music with the app (figure 5.27) or proceed directly to the *Ocarina* interface. The interface consists of four separate holes and an antenna icon at the bottom of the screen (figure 5.28). To play musical notes, users simply blow into the iPhone microphone while covering combinations of holes with their fingers. The sounding notes all correspond to a musical scale, which is chosen by the player. Tilting the phone downward while blowing into the microphone adjusts the vibrato rate and depth of the sounding note. Together, these affordances abstract the nuances of playing an actual ocarina—breath control, understanding musical scales, and producing vibrato with fingers rather than digital accelerometers—into an easy-to-use app.

Like other apps described in this chapter, *Ocarina* incorporates social-media sharing capabilities. By tapping the antenna icon at the bottom of the interface, the user is taken to a 3D map of the world that displays bright lights and plays sounds from locations where other *Ocarina* users are creating music (figure 5.29). Double-tapping locations on the map allows the player to "zoom in" on specific regions and listen to melodies being played in real time by *Ocarina* users in those areas. For musicians, the usability of the *Ocarina* interface combines with the global networking capabilities of the iPhone to align the ubiquitous production of the app with the mobile affordances of instruments like the harmonica. For casual smartphone users,

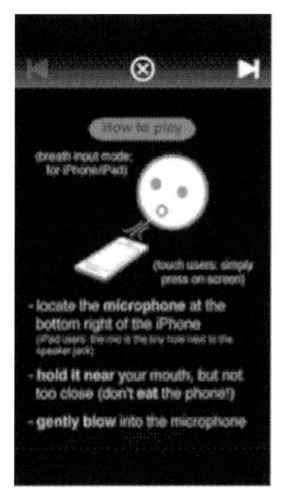

Figure 5.27 *Ocarina* tutorial (screenshot taken by author).

the app transforms the iPhone from an everyday tool into a creative platform for individual and collective expression.

On a technical level, *Ocarina* established many of the iPhone app design affordances discussed earlier. The app employs a graphical design that blends familiar visual and haptic metaphors from existing instruments with abstract patterns and gestural affordances unique to the smartphone. Further, the app embraces the ubiquitous computing design philosophy established by Weiser. App designer Wang discusses how the app shifts the "typical screen-based interaction to a physical interaction, in our corporeal world," where

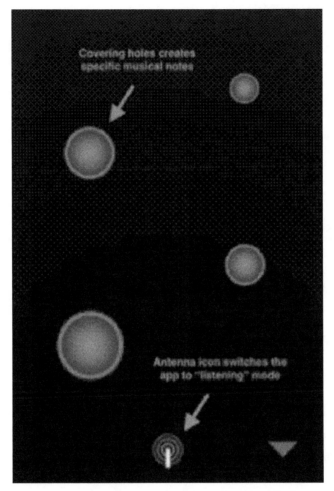

Figure 5.28 *Ocarina* main interface (screenshot taken by author).

the user engages the device with "palpable dimensions of breath, touch, and tilt."[57] Additionally, social-media integration and the ability to upload user-generated content to the cloud promotes the development of user communities surrounding the app, further enhancing the perception of intuitive, direct manipulation in the corporeal use of the device.

On a broader level, the generative nature of the app epitomizes a shift in perceptions of technology and cultural value, from one that privileges technical skill to one that embraces a disintegrating dichotomy between expert and nonexpert users. As a result of the democratization of gestures, design

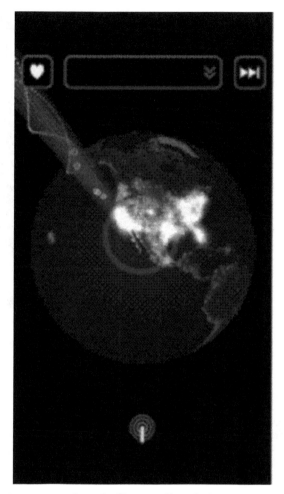

Figure 5.29 *Ocarina* social-media "listening" mode (screenshot taken by author).

affordances, and social connectivity in mobile media production and consumption, music making is valued for the experiences it generates rather than the sounds. For example, an American soldier in Iraq describes *Ocarina* as "my peace on earth. The globe feature that lets you hear everybody else in the world playing is the most calming art I have ever been introduced to." For this user, the values of global interconnectivity inherent in social and mobile media are embedded within the design of the app, which is perceived to "bring the entire world together without politics or war."[58] Here producing musical sounds is less important than the experience of connecting with other *Ocarina* users from around the world.

At the same time, it's hard to dismiss how users become addicted to their smartphones, their eyes glued to the glass screen as they attempt to drown out the world around them. While developers promote user-generated content as a positive force that encourages creativity and the open sharing of ideas across a social network, others view it as a form of free labor generated for the profit of the developers. In the same way that app design hooks users in by focusing the physical and cognitive capabilities of the user, the free promotion and cloud-based content development garnered by the iPhone exemplify the type of "immaterial labor" captured by what many have termed "cognitive capitalism."[59] In this situation, control over an intellectual workforce increasingly withdraws from direct management of the production process and instead exercises "an indirect power, based on ownership of intellectual assets, in the way a landowner might extract revenues from tenants of a piece of land."[60] Here the aesthetics of app design align with the economics of software development and distribution, as the structures through which design and economics function are hidden from the user. Exposing the factory labor of women building electronics used in sound technologies, for example, musicologist Lucie Vágnerová writes, "The notion of music technologies' democratising powers is a myth and a miscalculation: the tropes of broad access and effortless music-making feed off of labour happening elsewhere."[61]

Ultimately, the iPhone is both a generative creative platform and a tool for the proliferation of controlled consumption in twenty-first-century capitalism. Music making, for iPhone users, occurs not in the realm of sound production itself but rather in the perceptibly democratizing affordances of the device, as well as in the generative social and consumer platform the iPhone has become. The unique aspect of Apple's upgrade culture lies not in how it creates polarizing debates between technological solutionists and Luddites but in how it turns the acts of sociocultural negotiation at the heart of all technological change into dialogic experiences. As new apps are developed every minute and new iPhone hardware is rolled out on an annual basis, nonexpert musicians and everyday users of mobile media continue to experiment with music making, often for the first time in their lives. In contrast to the controllerists described in chapter 4, these individuals aren't concerned with distinguishing themselves as "musicians" but rather embrace their dual status as musical producers and media consumers.

The major innovation of mobile media design lay in its ability to embed the creative processes of music making into the everyday life of the consumer, integrating the previously separate venues of work, leisure, and personal well-being. Just as the multitasking capabilities afforded by iPhone design have resulted in the integration of various forms of mundane labor, companies such as Apple are increasing the extent to which consumers rely on mobile technologies in order to fulfill these everyday practices. At the same time, it's hard to dismiss the sense of joy felt when someone with no traditional musical training creates music for the first time. As much as the iPhone turns the process of music making into a form of labor to generate profit for Apple, the experimental design of many music-making apps reveals the importance of shared play and exploration at the heart of all musical experiences, regardless of technical skill.

PART III
SOFTWARE AS GRADUAL PROCESS

6

Worlds of Sound

Following the growth of low-cost internet game distribution platforms such as *Steam*, "indie" became a widespread descriptor for a gaming culture made up of developers, players, and distributors interested in creating an alternative to what was rejected as a glutted, uninspired, and corporatized game market. The rise of the indie aesthetic has followed a number of parallel developments in the gaming world, including the proliferation of gestural controllers such as the Wiimote and Xbox Kinect,[1] open-world "sandbox"-style games such as the popular *Minecraft* (2009), an ever-expanding market for "casual" mobile gaming,[2] and a growing community of amateur game developers and distributors. As a set of technical practices among game designers, the indie development community has foregrounded key features of contemporary games, including dynamic ambient audio, computer-generated simulations of natural ecologies, and a focus on game mechanics over visual fidelity and narrative complexity.[3] Together, these cultural, aesthetic, and technological shifts typify how alternative approaches to game design deploy digital aesthetics typical of large-scale studio development while also attempting to differentiate their small-scale offerings from those of major software companies.[4]

Through analyses of sound design in games such as *Fract OSC* (2014) and *Proteus* (2013), this chapter aligns the aesthetics and creative practices of game designers in the indie gaming community with traditions of generative art and music.[5] In doing so, I outline a broader musical turn that has occurred within gaming culture—one in which sound becomes a primary mechanic in the shaping of generative aesthetics at the heart of multiple forms of digital art. While new media or "post-media" aesthetics often prioritize continuity or rupture in relation to the histories of cinema,[6] literature,[7] or fine art,[8] the contemporary practices and historical precedents of generative aesthetics ground an understanding of the gaming experience as establishing both a historical lineage with earlier forms of process-oriented computer art and an increased focus on sound as a key medium for design and creativity in interactive media.

Push. Mike D'Errico, Oxford University Press. © Oxford University Press 2022.
DOI: 10.1093/oso/9780190943301.003.0007

Sound designers embed generative mechanics within a game's sound engine, which, rather than being a "hidden" part of the software code, manifests in practices shared between indie game players and generative music composers. Instead of perceiving the technical structure or code of the game as determining form, effects, and reception, as a platform-oriented study might do,[9] locating the sound engine in this larger history of process-oriented practices and forms allows us to understand how sound in digital media is not simply a fixed background effect but also a key element in the audiovisual interface that's generated through the process of HCI. Further, framing game software as a manifestation of a larger generative aesthetics suggests a new model for sound-oriented gaming experiences.

The game industry in the early twenty-first century has been culturally and economically divided between indie game developers on one hand and Triple-A games on the other. Technically, the term "Triple-A" refers to a grading scale used to denote production value in the games industry, with AAA being the highest rating. On a practical level, though, this distinction is rooted in the size of the budget and production team dedicated to the release of a given game. Indie game development often consists of small teams, sometimes even a sole individual with a meager budget. In contrast, Triple-A development is carried out by multiple creative teams housed within large entertainment companies with immense budgets. As a result of the considerable resources poured into Triple-A production, these games are sometimes categorized as blockbusters—a classification justified by the often-massive opening-weekend sales.[10]

In addition to economic disparities, the indie versus Triple-A dichotomy can be characterized by contrasting audiovisual aesthetics. Differing ideas about the role of sound in games, for example, has often grounded the split between the two cultures. Following the blockbuster analogy, Triple-A games typically align their aesthetics with those of popular American action cinema. In first-person shooter franchises such as *Call of Duty*, sounds typically enhance the effects of visual cues, serving as sonic amplifiers of the (mostly) male action hero at the edge. Assault rifle barrages are echoed by quick rhythmic bass and percussion chops, while the visceral contact of pistol whips and lobbed grenades marks ruptures in time and space as slow-motion frame rates mirror bass drops in sonic texture and rhythmic pacing. "Hardness" is the overriding affect; compressed, gated kick and snare drum samples combine with coagulated basslines made up of multiple oscillators vibrating at broad frequency ranges, colonizing the soundscape by filling every chasm of the frequency spectrum. In this example, sound is the affective catalyst for the emanation of an unabashedly assertive, physically domineering, and adrenaline-addicted masculine culture that has defined a large portion of Triple-A games.[11]

The aesthetics of indie games are subtler, employing minimalist audiovisual aesthetics reflective of the small-scale nature of independent game development. In 2009, indie development company Mojang released its sandbox survival classic, *Minecraft*. In contrast to the cinematic graphic fidelity and epic action sequences of many blockbuster first-person shooter games, *Minecraft* drops the player into a randomly generated world of pixelated landscapes, with no resources and few instructions on how to proceed. As the player learns the mechanics of the game through almost completely unfettered exploration, an entire ecology of plants, animals, and weather patterns seems to emerge spontaneously, resulting in what feels like an endless world of lo-fi abstraction. Ambient sounds and peaceful piano melodies fade in and out, providing the soundtrack to everything from planting seeds in a garden to watching the sunset. By providing subtle cues that help guide the player through the game, music functions like any other building block in the virtual sandbox—as architecture, landscape, and built environment. ▶

The refreshing feeling of nonlinear exploration afforded by the open world of *Minecraft* appealed to a generation of gamers seeking to escape the seemingly rigid gameplay of many Triple-A releases. Within only five years, *Minecraft* became one of the top three highest-grossing games of all time, selling more than 54 million copies and influencing a generation of indie game designers and developers.[12] Moreover, it defined many features of subsequent video games, including a focus on underlying game systems and mechanics rather than cinematic display resolution and processing power, as well as the DIY entrepreneurial spirit of indie culture more broadly. Designer Craig Stern captures the oppositional attitude well, describing indie as an aesthetic that "empowers the smallest and the freest of game developers. It is a declaration of creative freedom."[13]

Inseparable from the indie aesthetic are the politics of the independent games community, which are equally concerned with empowerment and equity. Contrary to conventional wisdom about gaming being a male-dominated activity, women were huge players in arcade culture and early gaming communities in the late 1970s and the '80s. As design researcher Gabriela T. Richard and media theorist Kishonna L. Gray note, two parallel developments ended up turning gaming into a "guys thing": first, "once games invaded the home and could be found in bars in the late 80s and early 90s, they started to gain a stronger male following," and second, "more marked gendered and racialized representations emerged in games once graphics began to improve."[14] These gendered connotations began to shift with the emergence of indie games. Specifically, in 2014, Eron Gjoni—ex-boyfriend of game developer Zoë Quinn—wrote a negative blog post about Quinn which resulted in internet users falsely accusing Quinn of having an unethical relationship with video game journalist Nathan Grayson. The post sparked an online harassment campaign from so-called Gamergaters that included doxing (releasing personal information), rape threats, and death threats against Quinn, fellow game developer Brianna Wu, and media critic Anita Sarkeesian. In what became known as the Gamergate controversy, the stakes of the 2010s culture wars were made clear as the game industry was forced to acknowledge its severe lack of representation from women and people of color.

While the Triple-A versus indie debate isn't necessarily black-and-white, the indie games community has made clear attempts to alleviate the industry problems that Gamergate made apparent. The Los Angeles Indiecade Festival continues to program panels on social justice topics from black cyberfeminism to being an LGBTQ+ ally, and the Game Developers Conference hosts an entire track of sessions on advocacy that addresses issues of diversity in character representation, accessibility, mental health, and social impact. Indie's aesthetics and politics are deeply intertwined both explicitly and implicitly, reflected in the personal values of the developers, the back-end design of the gaming mechanics, and the aesthetic design of the interfaces used to interact with the game. As I argue throughout this chapter, these procedural interfaces blur the boundary between player and developer, game and reality, aesthetics and politics.

The indie attitude of embracing limited resources and self-consciously attempting to assert oppositional practices in game design has encouraged the development of highly experimental games, often conceptualized as sandbox spaces designed for open-world exploration rather than linear narrative. Developer Jonathan Mak describes his musical puzzle platformer *Sound Shapes* as "walking through a world of music."[15] In describing *Fez*, the vast "2D platformer set in a 3D world," designer Phil Fish states, "FEZ aims to create a non-threatening world rich with ambiance, a pleasant place to spend time in."[16] Games such as *Osmos* and *Splice* position the player within extreme macro and micro ecologies, from cosmic space to microbial DNA. These examples highlight the acute vulnerability and insignificance of the player, reflecting what Jack Halberstam calls the "queer art of failure."[17] While it's often assumed that winning generates the apex of pleasure in gaming, Halberstam thinks through how failure can also be a rewarding way to open up alternative possibilities of knowing and being in the world. Unlike many Triple-A games that thrive on immediate gratification and maximalist gameplay mechanics, indie games complicate the meaning of success, thus "queering" the normative rules and goals of society. Arguably, one of the core sources of pleasure in indie gameplay comes from how the games subvert the neoliberal impulse to maximize productivity by teaching us how to be content even when we're doing nothing at all. Procedural listening, in this context, is a way of listening with the game, hearing more fully not only its rules and goals but the social and political context beyond the screen.

While game developers may never fully agree on what is or is not "independent" or "mainstream," I build my argument on the position that the audiovisual aesthetics of indie games are defined first and foremost by the presence of what artists have called "emergent aesthetics."[18] In contrast to aesthetic theories of the eighteenth and nineteenth centuries, which attempted to create rules to explain things already present in the world, emergent aesthetics attempt to create rules that, in turn, generate the world itself.[19] In *Minecraft* and the other games previously mentioned, emergent aesthetics enhance the player's experiential engagement with the virtual world, as the visual environment is seen to materialize in the process of gameplay. In this context, sound occupies a dual position. First, audio *content* arises in a seemingly spontaneous manner, as a result of a feedback loop between the player's actions and randomizing algorithms in the game's software code. Second, the *processes* and rules of gameplay are often defined by emergent mechanics ingrained within the game's sound design. Emergent aesthetics in sound thus allow the game environment to become an interactive digital instrument of sorts. Rather than pursuing a set of goals toward a linear narrative with a precomposed soundtrack, emergent aesthetics encourage the player to interact critically with the mechanics of the game itself through forms of procedural listening.

"Explore an abstract broken-down world built on sound. Rebuild its structures and forgotten machinery. Create your own sounds and music within the world."[20] This game synopsis from the developers' website is the only set of instructions you receive before entering the cold, dark world of

Figure 6.1 The screen upon entering the initially bleak world of *Fract OSC*
Source: FRACTgame, Flickr account, March 1, 2014, https://www.flickr.com/photos/fractgame/12863049813/.

Phosfiend Systems' *Fract OSC*—a music-based puzzle game released on both Mac and Windows operating systems in April 2014. Like many indie games, *Fract* basically consists of a 3D, open-world environment that privileges rules and mechanics over visual fidelity. Upon launching the game, you're placed on an unlit platform of black polygons and presented with what appears to be a white neon elevator in front of you and a spiral ramp just beyond your purview (figure 6.1).

As you ascend the ramp, you're confronted with an enclosed structure of locked modules only later revealed to be components of a real-time subtractive synthesizer. There are few resources at your disposal, virtually no in-game hints on how to proceed, and a limited GUI. As you traverse this sparse yet monumental world of sci-fi abstraction, you're introduced to the simulated physics of the visual space *through* an exploration of sound.

The visual environment of *Fract* consists of abstract fragments of actual synthesizer components—buttons, knobs, sliders—that function as sonic and spatial markers, allowing the player both to create original sounds through the game's built-in subtractive synthesis engine and to navigate the virtual world through warp points (figure 6.2). Meanwhile, the precomposed soundtrack evolves with the player's understanding of the game mechanics, starting from static bass drones and subtle synth pads to vibrant arpeggios and immense lead melodies that mirror the towering structures and radiant

Figure 6.2 In the guts of the synthesizer; the bass synth module being constructed in-game

Source: FRACTgame, Flickr account, February 11, 2013, https://www.flickr.com/photos/fractgame/8466277208/.

lighting schemes gradually generated throughout. While there are no clearly defined goals in *Fract*, the primary purpose of the game is to gradually reconstruct and revive a defunct world through an exploration of its physical and sonic possibilities. Sound is the primary building block of gameplay and the fundamental mediator between game and player. ⏵

The emergent aesthetics of *Fract* evoke the exploratory dynamics of early puzzle games such as *Myst* (1993) while echoing the gameplay of rhythm games such as *Frequency* (2001). Each of these games results in a nonlinear experience based on exploration through a world whose mechanics are defined by its sonic structure and the gestural articulation of that structure by the player. The open-ended design of *Fract* highlights broader emergent aesthetics that have influenced the DIY attitude of indie game design and development, embracing limited resources to accomplish seemingly unlimited aims. Emergence is noticeable in both the game design and the player experience, as the audiovisual environment is seen to emerge literally in the process of gameplay.

Ed Key and David Kanaga's *Proteus*—a self-proclaimed "game of audiovisual exploration and discovery"—is another example of a game that uses emergent aesthetics to provide an alternative experience to blockbuster Triple-A titles.[21] *Proteus* was released as a ten-dollar digital download for

Mac and PC computers in 2013, and its pixelated 3D world is designed in such an open-ended way that nearly every element of the game is built using design principles intended to facilitate the perception of emergence in the player experience. Each time the game loads, a sparse island landscape is procedurally generated based on the Global Positioning System location of the user, resulting in the creation of a unique world each time it's played. There are no strict rules to follow, and players are given no specific controls to navigate the game. They are simply dropped onto an island, free to observe the world unfolding before them, emphasizing a typical focus of the indie aesthetic on the system rather than the player or designer.

The sound of *Proteus* is generated from a combination of ambient noises and synthesized pitches produced from every aspect of the visual environment; it can be triggered based on the player's proximity to specific objects in the world and varies in frequency and timbre depending on how the player interacts with the objects. For example, walking toward an animal causes it to scurry away, producing rapid, high-frequency string plucks in its wake. The longer the player explores a given interaction, the richer and more complex the sonic result. Loitering in a small forest for a long enough period of time, the player will hear a single synthesizer pad tone develop into an immense ambient soundscape in parallel with an increasing number of leaves falling from the trees. Sound acts as one of the few sensorial guides through the open world of *Proteus,* intensifying the emergent aesthetics that define the visual landscape (figure 6.3). The gameplay experience is process-oriented, experiential, and improvisational, as the mechanics of game design work in conjunction with player experience, and the seemingly random juxtaposition of sound and image creates the emergent aesthetic. ▶

Fract OSC and *Proteus* represent only a couple examples of how sound plays a crucial role in defining indie games' aesthetics and gameplay mechanics. In spite of this, conventional thinking about games has yet to address the shared relationships between *sound* design and *game* design. Media theorists tend to understand sound in gaming as a matter of either the subjective experience of the player or the structural or platform properties of the game as a designed system. Musicologists, for example, have provided useful insights into how the multisensory space of games influences the subjective and cognitive capacities of the player. Karen Collins introduces the idea of "kinesonic synchresis" as the ability of game sound to carry "connotations of its haptic and visual associations even in the absence of these modalities in media."[22] In this context, player interaction with sonic material means

Figure 6.3 A typical afternoon in *Proteus*
Source: https://store.steampowered.com/app/219680/Proteus/.

that sound becomes a sensory guide for players as they navigate the virtual world, as in *Fract*. William Cheng and Kiri Miller think through how players construct ethical subject positions in response to virtual actions committed through forms of embodied interaction in games such as *Fallout* and *Guitar Hero*.[23] While these thinkers, among others, have done excellent work in bringing the experience of sound to the forefront of video game studies, their focus generally remains on the subjectivity of the player rather than the technical work of game and sound designers.

At the same time, research from sound *designers* and game audio *engineers* has provided practical knowledge about how sound can be structured technically to create emergent experiences for the player, thus foregrounding the technical systems of the game over the aesthetic experiences of the player. The programming technique known as procedural content generation (PCG) has been a particular focus among developers interested in the "dynamic" and "adaptive" capabilities of audio in gameplay.[24] Since the commercial proliferation of games in the 1980s, PCG has facilitated emergent aesthetics in game design by generating content randomly or pseudo-randomly through algorithmic computational processes rather than through custom-modeled work by the game or sound designer. Collins describes Japanese composer and artist Toshio Iwai's *Otocky* (1987), a side-scrolling shooter game in which the sound of the player's firing rhythmically quantizes in real time, becoming the melodic accompaniment to a two-note bassline.[25] Moreover, Collins and others have highlighted how PCG gave early game developers an efficient strategy for preserving computer memory and storage, thus foreshadowing the indie aesthetic of embracing limited resources in game development. *Times of Lore* (1988), for example, used random-number generators to give the impression of variety despite a limited number of preprogrammed musical patterns.[26] Significantly, PCG is agnostic to media or content type, generating anything from levels and physical assets to UI elements and sound design. As a result, PCG is now a common technique for creating digital content among indie game developers as well as artists across media platforms.

Game theorists have attempted to align the *technical* practice of PCG with broader *aesthetic* values in the culture of indie gaming. Ian Bogost characterizes the indie aesthetic through what he calls the "proceduralist style" of game design, highlighting design trends such as an orientation toward process and introspection, a foregrounding of game rules and mechanics, and an abstraction of visual content (rather than verisimilitude).[27] Returning to the example of *Minecraft*, we can see how this idea plays out. The low visual fidelity of the game world evokes the retro aesthetic of vintage eight-bit game consoles, in which graphical elements were built by piecing pixels together one by one. The pixelated visual aesthetic of the game mirrors the primary game mechanic in *Minecraft*, which involves piecing together actual blocks of raw materials in order to build functional objects. The abstraction of visual content in both visual design and gameplay encourages players to focus on their role as a builder, logically piecing together distinct combinations of blocks in order to build shelter, cook food, farm crops, and more. Player introspection is encouraged, because no preset rules are established for the relationships the player chooses to foster in the game. The vast scale of the emergent worlds in *Minecraft* facilitates both completely hermetic existences and the proliferation of thriving civilizations. Just like the virtual construction of the game's terrain through PCG, learning how to play *Minecraft* is an emergent process. These procedural elements are most noticeable in the sound of indie games, which functions as an aural guide through the mechanics of the game.

Emergent aesthetics in indie games extends a long and diverse tradition of generative aesthetics in media art since the 1960s. Tracing this history allows us to uncover shared relationships between computational media platforms in the twentieth century but also to make more explicit connections between the emergent aesthetics of musical composition and gaming. Broadly defined, generative aesthetics focus on the relinquishment of creative control in the creation of artistic content, instead privileging how aesthetic effects and values are embedded within the technical structure of computational systems themselves. In his overview of visual artists working with computers from the 1960s and '70s, Frank Dietrich notes that many artists shied away from standard forms of human aesthetics and creativity in favor of the computer's "nonhumanness," which was seen to facilitate experiments in random image generation and algorithmic drawing.[28] Frieder Nake's *Matrix Multiplications*, for example, converts a mathematical process—a matrix successively being multiplied by itself—into an interval-based grid. Like many early experiments with algorithmically generated art, the symmetrical yet seemingly random layout of *Matrix Multiplications* visually depicts the process of its own making. The perceived stoicism of the computer is pitted against the conventional standards of beauty imbued by the "natural" world.

Essentially, generative art is about creating processes that, in turn, create artworks. Artist and designer Mitchell Whitelaw writes that in generative art, "the work is entirely shaped by the construction of its underlying system, its configuration of entities and relations."[29] In focusing on the materiality of the creative medium as an engine for the creation of art, generative artists often embed their aesthetic values within the technical structure of artworks themselves—a phenomenon shared by the use of PCG in game design. For example, the multimedia studio and design group Universal Everything specializes in the augmentation of public space through the creation of "digital motion canvases" and dynamic "video walls" that respond and react to the physical bodies that inhabit them. In its 2014 "Infinity Room" video installation created for Microsoft, image repetition and recursion are combined with a constantly evolving visual backdrop to highlight the ubiquitous presence and fleeting nature of digital data. In the South Korean "Hyundai Vision Hall" video wall, computer-generated visual noise provides abstract simulations of seemingly "natural" phenomena such as bubbling lava and crashing waves, spread across 720 microtiles on a twenty-five-meter-wide, four-meter-high wall, surrounded by thirty-six-channel surround sound. These virtual spaces exemplify how generative art is shaped by a process-oriented relationship between the design of a technical system and the feedback of its users.

The skills fostered in creating generative art and design of these sorts—logic, computer programming, algorithmic thinking—contrast strikingly with the standard "fidelity" model of artistic design which values faithful reproduction of "real" objects and textures. Rather, concepts surrounding generative aesthetics are more closely aligned to those found in historical discussions of AI and computer science, focusing on the construction of iterative systems capable of self-generating content "organically."[30] To return to the *Minecraft* example, Microsoft began testing AI platforms in the spring of 2016 in order to "help agents sense and act" within the game.[31] The indie games movement has embraced the perceived "organicism" of generative aesthetics as an alternative to the proprietary, large-scale nature of blockbuster game development—a cultural stance shared by other forms of digital art.

In addition to its use in computational arts, the generative aesthetic is noticeable in the experimental and electronic music of the mid- to late twentieth century. Deriving from compositional theories of Darmstadt composers such as Karlheinz Stockhausen and Gottfried Michael Koenig, the indeterminate music of John Cage, and the gradual processes of Steve Reich, "generative music" describes responses to compositional methods based on the intentional control of musical elements, instead focusing on self-generating compositional systems.[32] While these composers applied computational concepts and procedures to musical performance, early computer music pioneers embedded generative aesthetics within computational systems themselves. In 1957, Max Mathews wrote *MUSIC*, the first computer program capable of generating digital audio waveforms, as well as one of the first widely used programs for making music on a computer.[33]

You might recall that software developer Miller Puckette wrote *Max* in the mid-1980s as an homage to Mathews, a program that by the 1990s became ubiquitous among musicians and artists seeking to create generative music in real time. *Max* is based on principles rooted in the procedural nature of computer programming.[34] The work of Mathews, Puckette, and other computer musicians interested in algorithmic, generative, and "machine" processes highlights the significance of AI and "organic" emergence in musical composition. With the rise of the "procedural style" among game designers, the core techniques of AI and emergence would also be embraced by sound designers.

Having traced seemingly disparate histories of emergent aesthetics in indie games and generative aesthetics in twentieth-century artistic movements, I offer the concept of emergence as an overarching model for understanding the interconnected relationships between the design of interactive systems and the experience of HCI in sound, games, and software. The concept of emergence has often been used to talk about how the technical development and social effects of technology can be understood in nonlinear ways.[35] Social constructionist approaches like those of Bruno Latour understand the existence of sociotechnical systems as preconditions for the emergence of more complex sociotechnical systems.[36] According to Latour's "Actor-Network-Theory," both technology and society are seen as dynamic intermediaries that emerge throughout a broader history of sociotechnical evolution. For example, media theorist Sandy Stone describes the ways in which multiplicities of identities are capable of emerging at the interface between socially constructed bodies and "virtual" technologies.[37]

Among professional designers, emergence has been used as a framework for integrating the sociocultural factors of UX with the technical affordances of specific products. Interaction designers talk about the idea of "emergent interfaces" as a strategy in outlining a more integrated framework for feature development within a shared software environment, for example.[38] In the context of games, designer Penny Sweetser introduced the term "emergent gameplay" to describe how interactions between objects in the game world or the players' actions result in a second order of consequence that was not planned, or perhaps even predicted, by the game developers.[39] Together, the concepts of emergent interface and emergent gameplay highlight the interconnected relationships between the mechanics of sound design and player experience in indie games. Thinking through the concept of emergence from the perspectives of both social theorists and professional designers foregrounds how technical systems co-emerge with social structures and values, allowing us to consider the integrated and dynamic relationship between society, culture, and technology.

By expressing generative aesthetics through the technical structure of the interactive system, artists, musicians, and designers conceptualize gaming as an integrated media experience. In short, generative aesthetics and emergent gameplay coalesce to form what I call a procedural interface. In chapter 3, I introduced the concept of procedural listening to describe the practice in which computer users focus their attention on the process-oriented mechanics at play in the inner workings of software itself, rather than the content or narrative being enacted by the interaction. While the term "interface" most commonly evokes the GUI of computer operating systems, or more generally the control surface of a given object, I use the term "procedural interface" to describe the overall experience enacted by the sociotechnical system making up the media platform, including the values and aesthetics designed into its technical infrastructures.[40] Procedural interfaces provide bridges between the mechanics of rule-based systems and the aesthetic experience of the musician or player in response to those systems.

Considering interfaces as entire sociotechnical systems, we can think about how both the technical structure and the user's experience of the software emerge in a seemingly spontaneous manner. As the user interacts with a procedural interface, the mechanics of the system are gradually made apparent to the interactor, and it's this decoding of the system's rules that becomes the primary function of the experience. This concentrated attention to the dynamics of technological interaction is apparent in a range of media arts, including computer-generated visual art, generative music, and indie games. In each case, the morphology of digital sound, modeled from the seemingly "organic" principles of artificial intelligence and real-time synthesis, functions as a sensory guide in the experience of emergence. The idea of procedural interfaces pushes beyond existing conceptions of generative aesthetics and emergent gameplay by focusing on the space in between process (gameplay) and product (game), foregrounding how a user comes to know, interact with, and embody a sociotechnical system through aesthetic ludic practices.

Fract OSC is a paradigmatic procedural interface. The process of foregrounding the mechanics of sonic interaction is most apparent in *Fract's* "Studio Mode." As the player enters a small portal-like enclosure near the entrance of the main game world, the dark landscape of *Fract* is brightened by a transparent panel containing a colorful grid. Although there are no textual instructions or visual cues for how to proceed, players familiar with electronic music composition recognize the abstract knobs and virtual buttons as simulations of graphical interface elements typical of DAWs (figures 6.4 and 6.5). As players tinker with the transparent neon LCD display in front of them, sounds gradually emerge, morph, and fade away. *Fract* is not simply a game in which players explore the inner structures of a synthesizer, but it's also a compositional environment for creating original music. The game serves as what artist Norbert Herber calls a "composition-instrument . . . a work that can play and be played simultaneously," resulting in a form of emergent gameplay that represents "a kind of becoming where myriad events collide to unfold as experience in the course of their play."[41] Here the techniques and processes of *audio* production—rather than 2D or 3D visual cues—guide the fundamental mechanics of gameplay.

"Studio Mode" is an apt component of what makes *Fract* a procedural interface, in that it's an *aesthetic* element of gameplay that reveals the *technical*

Figure 6.4 *Fract OSC* "Studio" sequencer
Source: https://cultmtl.com/2014/04/the-incredible-music-machine/.

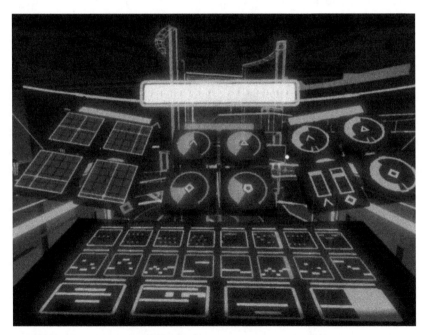

Figure 6.5 *Fract OSC* "Studio" editor
Source: https://techgage.com/news/fract-osc-a-puzzling-sequencer/.

structure of the game's sound engine. The *Fract* sound engine is a mediator in the creation of emergent gameplay, both designed using principles of PCG and shaped by the compositional choices of the player. As players tinker with music composition and audio synthesis in "Studio Mode," they simultaneously set the sound engine in motion and are made aware of its technical structure. Like many generative art forms, *Fract* is self-referential and celebratory in regard to the processes of its own construction.

Developed by programmer Henk Boom, the *Fract* sound engine is constructed using the generative music and synthesis capabilities of *Pure Data (Pd)*—the open-source instance of the popular *Max* visual programming "environment," initially developed by Puckette.[42] Similar to the cultural position of indie games themselves, *Pd* is the epitome of alternative software for audio production. Puckette originally designed the program to liberate musicians from what he perceived to be creatively limiting features of commercial DAWs. Most commonly, audio production software conceptualizes sound as recorded material that's temporally mapped onto crucial points of what's essentially a linear narrative structure.[43] In contrast, *Pd* encourages the composition of generative sound by allowing the designer to relate digital sound processing functions ("objects") to one another using virtual patch cords, resulting in process-oriented musical composition systems that respond to user input in real time (figure 6.6).

The object-oriented nature of *Pd* is highlighted by the fact that it's a "visual programming language" that allows users to create programs by manipulating graphical elements of the interface, rather than using text-based code to specify commands. Since the development of the first computer languages in the 1970s, one of the main goals of visual programming has been to make the practice of coding easier to learn.[44] *Scratch*, the most popular visual programming language, is designed specifically to teach children how to code, organizing graphical elements as colorful blocks that fit together like Lego pieces. Echoing the core elements of all procedural interfaces, visual programming languages such as *Pd* and *Scratch* make apparent to the user the rules of the software through graphical and audio cues. In a similar way, indie games such as *Minecraft* and *Fract OSC* employ visual programming principles in the structure of both their technical design (using a visual programming language to build the software's sound engine) and gameplay (the player arranges graphical elements to build the virtual world as he or she experiences it). The resulting interface is procedural in that both the

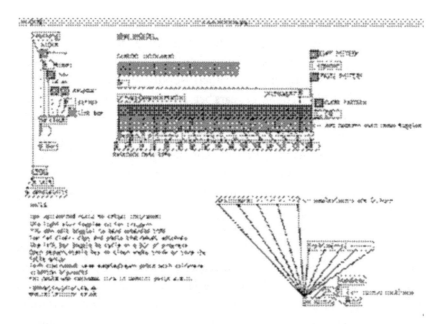

Figure 6.6 "PD Drum Machine" by Nullpointer, in the style of a Roland TR-808

Source: Nullpointer, accessed August 28, 2020, http://www.nullpointer.co.uk/content/pure-data-patches/.

technical code and the aesthetic audiovisual content are procedurally gener-ated in real time.

Boom integrates the modular sound engine of *Fract* with the 3D, gen-erative visual aesthetic of the game. In order to combine disparate sound and game software, he uses *libpd*—a collection of resources that turns *Pd* into an embeddable audio synthesis library. While *Pd* is a stand-alone soft-ware program that runs in a desktop computer environment, *libpd* allows programmers to embed the work they create with *Pd* into other software. Most commonly, the library has been used by iOS developers looking to in-tegrate the generative music capabilities of *Pd* with mobile app development software. For example, the developers of the 2008 iPhone app *RjDj* used *libpd* to create what they referred to as "reactive music" programs that manipulated preexisting musical content in various ways based on real-time user interac-tion. The default *RjDj* music player receives ambient noise from the listener's acoustic environment, procedurally remixing and distorting the content as iPhone users go about their day.[45] The proliferation of generative music apps parallels the rise of indie games, aligning the vernacular creativity imbued by mobile media with the ludic affordances of procedural interfaces.

Specifically, Boom uses *libpd* to integrate *Pd* with *Unity*, a game engine especially popular among indie developers working with 3D virtual worlds. Released in 2005, just a couple of years before the launch of the first iPhone, *Unity* prides itself on its interoperability across media platforms (figure 6.7). As the name suggests, *Unity* is a cross-platform engine used to develop games for a range of media, including personal computers, gaming consoles, mobile devices, and websites. Examples of games created using *Unity* include iOS puzzle apps *Prune* and *The Room Three*, open-world console games *Firewatch* and *Rust*, and VR "experiences" *Job Simulator* and *Tea Party Simulator 2015*. These games share core elements of the previously outlined procedural style in indie games, including an orientation toward process and introspection, a foregrounding of game rules and mechanics, and an abstraction of visual content (rather than verisimilitude). The generative audio capabilities of *Pd* and the adaptability of *Unity* combine to form the procedural interface of *Fract* as a whole—a game in which audiovisual content, as well as technical infrastructure, is defined by a real-time, process-oriented dialogue between player and system.

Similar to the object-oriented interface of *Pd*, *Unity*'s visual programming environment became popular among developers for the efficiency with

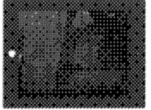

Figure 6.7 Promotional material highlighting the interoperability of *Unity* across twenty-one media platforms

Source: Unity3D, "Multiplatform," accessed May 2, 2016, https://unity3d.com/unity/multiplatform.

which one can make simple 3D game objects perform complex procedures. Game assets of various sorts are easily dragged and dropped into a 3D space, affording the user a "plug-in" style of modularity of layout similar to DAWs such as Ableton's *Live* (figure 6.8). For example, a virtual basketball object could be dropped into the 3D editor, which would consequently display a 3D basketball in the window. Then a physics simulator could be dropped onto the basketball object in the editor, causing the ball to move in a specific way. Finally, a sound sample of a bouncing basketball could be dropped onto the moving ball, which would cause the object to play back the sample every time it made contact with a solid surface. The instant feedback of the drag-and-drop editor conveys to designers the impression of direct manipulation—the feeling that they're directly interacting with the virtual world itself, rather than a programming software. In building the iOS puzzle game *Monument Valley*, lead developer Peter Pashley said he knew *Unity* would be the best software possible, because the game "is all about creating beautiful places, and for that you need artists to be able to work *directly with the world.*"[46] Just as the player of *Fract OSC* is placed in the position of audio designer, the software used to build the game positions the game designer in the role of a player, embracing ludic mechanics in an accessible manner to build the virtual environment. ⊙

Figure 6.8 *Unity* "Scene" view (screenshot taken by author).

The feeling of direct manipulation imbued by the visual programming environment not only makes visible to the user the rules of the programming language itself but also creates a more playful relationship between the abstract nature of coding and the aesthetic experience of gameplay. Game developer John Alvarado describes *Unity* as a "component-based game object system" in which the user can attach coded functions to any game asset, including inanimate materials, avatars, and sound. According to Alvarado, "It's real easy to add code components to any object you create in the game, whether it be a box you just made or an animated character. It's a very modular, object-oriented way of adding functionality to an object in the game."[47] In the same way that audiovisual characteristics and rules for interaction can be dragged and dropped onto game assets through *Unity's* 3D editor, so, too, can computer scripts be coded into the objects.

Let's return to the basketball example. Designers are not limited to the 3D drag-and-drop editor in terms of the types of relationships being created between the basketball and other game objects. If designers want to create an automated process in which the basketball bounces ten times every time it's dropped on a solid surface, they can code original rules for interaction using *MonoDevelop*—an integrated development environment (IDE) supplied with *Unity* that functions like most text editors for computer coding. The unique aspect of *MonoDevelop* is that it allows the designer to create custom behaviors for game assets that can be accessed from within the *Unity* editor. In this way, the abstraction of text-based computer code becomes externalized and made playable in the visual programming environment of *Unity*. While the editor itself may appear to be object-oriented in its layout, the primary goal of procedural interfaces in game design is to make explicit and tangible the previously implicit processes hidden behind software algorithms.

The process-oriented nature of the *Unity* editor is amplified by its capacity for rapid prototyping. In the same way that the editor blurs the line between visual programming and text-based coding, so, too, does the software blur the line between design and gameplay. As such, the creative workflow in *Unity* echoes the practices of musicians working with software such as *Live* and *Max*, both of which allow the creator varying degrees of rapid prototyping and modularity of interface layout.

As the interactive objects and rules of the virtual world are being constructed, the developer can switch between "Scene" (edit) and "Game" (play) modes, effectively alternating between the act of graphical programming and the experience of the code as it would appear to the player.

Designers can thus witness the effects of their programming choices in real time, without having to compile and run the game as a process separate from *Unity*. Mathieu Girard, CEO of Amplitude Studios, aligns the usability of visual programming languages such as *Scratch* with that of *Unity*, claiming that the greatest part of *Unity* is its workflow and "how easily you can create, edit, and integrate data and code. Building inspectors, editors . . . everything is *child's play*."[48] This ability to rapidly prototype the game world in real time is an experience shared by both designer and player, further highlighting the integrated technical and aesthetic affordances of *Fract* as a procedural interface. In effect, the design process becomes as experimental and improvisational as the experience of the game itself.

The integration of emergent gameplay mechanics through the use of scripts in *Unity* and generative music techniques through the incorporation of *Pd* distinguishes the sound design of *Fract* from traditional examples of procedural game sound. While procedural sound generation in games is often used to create musical variety or avoid bogging down the game's processor, the sounds produced as a result of player input in *Fract* are synthesized in real time by the game's sound engine (figure 6.9). In this way, the purpose of the sound engine aligns with the purpose of the game itself: to rebuild a defunct synthesizer by piecing together its core components. While a prerecorded soundtrack provides a sonic backdrop to many crucial moments in the simplistic narrative of *Fract*, sonic interaction design in the game is defined for the most part by a generative soundscape that emerges in the process of the player's dialogue with a constantly evolving audiovisual interface.

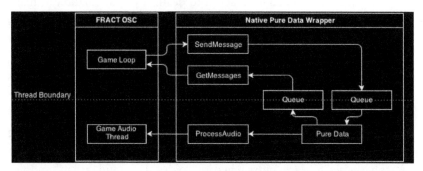

Figure 6.9 *Fract OSC* to *Pd* audio workflow, created by the game developers. Sounds emerge from a feedback loop between player interactions in the game (displayed on the left) and audio processing capabilities in *Pd* (displayed on the right)

Source: FRACTgame, Flickr, June 4, 2013.

Examining the contemporary gaming experience as a procedural interface made up of both emergent gameplay mechanics and generative sound design provides a useful bridge between player- and developer-oriented thinking about games and music. By building emergent aesthetics into both the technical structure of a game and the subjective experience of the player, it's possible to view the gaming experience as a sociotechnical system. This mode of thinking allows us to comprehend more distributed forms of agency in the experience of gameplay, dethroning the sound designer or composer as the sole factor in the creation of game sound. As examples of what Herber calls "composition-instruments," games such as *Fract OSC* position the player simultaneously in the roles of composer, designer, and experiential observer. Further, the close relationship between sound and image fostered by emergent aesthetics encourages sound designers and composers to broaden their knowledge of design and production techniques outside of strictly "musical" practices. As I've suggested throughout this book, the appearance of new interfaces for HCI always occurs in parallel with new conceptions of fundamental ideas regarding art and culture in the digital age.

Perhaps one of the lasting influences of the indie game movement is how it teaches players to slow down, look inward, and challenge themselves to grow. In his "Slow Gaming Manifesto," Artur Ganszyniec outlines the need within the industry for games that "challenge not my skills and reflexes but my assumptions and feelings . . . encourage me to take a break and come back later . . . let me experience them at my own pace . . . give me space to grow . . . [and] are slow."[49] Analogizing discriminatory design practices with a type of "technocorrections" that led to deaths by strangulation of many black people throughout the 2010s, Ruha Benjamin notes that the slowing down of design may offer a metaphorical and literal chance to breathe for people of color: "Forward movement, the ethos of design, matters, of course, but rarely does such an approach allow us to slow down and let ourselves breathe in ways that may be useful. . . . In the breathless race for newer, faster, better technology, what ways of thinking, being, and organizing social life are potentially snuffed out?"[50] Media theorists Anna Everett and S. Craig Watkins describe the "Racialized Pedagogical Zones (RPZs)" that video games have used to teach "entrenched ideologies of race and racism," and also how "gameplay's pleasure principles of mastery, winning, and skills development are often inextricably tied to and defined by familiar racial and ethnic stereotypes."[51] For an industry so deeply entrenched in the racist and sexist ideologies of the twenty-first century "culture wars," rethinking the goals of play may carry with it the potential to undo, at least partially, problematic assumptions and values that continue to push women and people of color out of the industry. More broadly, the idea of gamers accepting and even embracing "boredom" as a component of play contradicts the fundamental ideal of "immersion" in media design, instead encouraging the player—following designer Maisa Imamović—to eschew "traditions of thought and aesthetic practices that force him or her to think and create."[52]

The procedural interfaces of indie games are just one set of examples that highlight how artistic tools are becoming more abstract and complex, their design affordances and possible uses underdetermined and hence translatable across media formats and sites of technological production. As such, game developers and sound designers have begun to borrow software popular among computer musicians to facilitate emergent gameplay, just as DJs, VJs, and electronic musicians have incorporated software such as *Unity* in an effort to enhance the process-oriented aspects of their performances. Contemporary software for sound design and gaming continues to transform the creative possibilities of artists working within a larger historical trajectory of generative aesthetics and sociotechnical emergence. As game designers and digital musicians continue to explore the worlds of sound at the interface of culture and technology, these latent histories manifest in broader areas of cultural production.

7

Deep Listening

In 2016, Grammy Award–winning music producer Alex Da Kid teamed up with IBM's "Watson Beat"—a "cognitive machine" that generates entire musical compositions based on a few seconds of notes—to produce the song "Not Easy." Some may remember Watson from its 2011 appearance on *Jeopardy!* in which it defeated two of the show's most famous champions. Watson Beat, on the other hand, provides an example of how AI in the 2010s made bold moves into the world of art, design, fashion, and other forms of computational creativity. Computer engineer and IBM Research team member Janani Mukundan describes the simple process behind the machine: "To use Watson Beat, you simply provide up to ten seconds of MIDI music—maybe by plugging in a keyboard and playing a basic melody or set of chords—and tell the system what kind of mood you want the output to sound like. The neural network understands music theory and how emotions are connected to different musical elements, and then it takes your basic ideas and creates something completely new."[1] Behind the scenes, though, Watson is also surveying social-media posts and internet blogs to identify the most prominent themes from a given era and aligning these emotions with the lyrics of chart-topping songs from the same time. In deciphering the "emotional temperature" of popular music, Watson can create, according to IBM, "music that 'listens' to its audience."[2]

In yet another example in the long lineage of emergent aesthetics from chapter 6, Watson Beat revives the age-old question of computational creativity for the twenty-first century. Is it possible for AI to compose compelling music? That's the more familiar question. But what about the higher-level parameters of music making that are socially and culturally contingent? In the age of "big data" and "deep learning," we might ask, is it possible for AI to compose compelling music in the style of, and in response to, actual social and cultural products? More fundamentally, what might it mean socially and culturally for these "intelligent" machines to participate in the type of deep learning and deep *listening* practiced by tools such as Watson Beat?

Push. Mike D'Errico, Oxford University Press. © Oxford University Press 2022.
DOI: 10.1093/oso/9780190943301.003.0008

The term "deep listening" was coined by composer Pauline Oliveros in 1989 to describe a practice of radical attentiveness, highlighting the listening experience as an empathetic act that allowed listeners to feel more aware of themselves in relation to the collective universe. Here I'm using "deep listening" as a juxtaposition of two seemingly opposed definitions. First is Oliveros's use of the term, which is often seen as an attempt to escape the distraction and saturation of contemporary life. Second, I'm riffing on developments in the application of deep learning to music software—that is, the development of machine algorithms that analyze a massive corpus of musical recordings in an attempt to listen with the ears of a music consumer and to attune themselves to the "emotional temperature" of a time period. Techniques of AI simultaneously encourage computational forms of Oliveros's deep listening, while reifying the panoptical surveillance and algorithmic rigidity of machine listening. In other words, I argue that deep listening in the twenty-first century is being practiced by both machines and humans, for better or worse.

These questions and tensions have been taken up by the major "cognitive machines" dedicated to art and culture in the 2010s, including Google's Magenta, Sony's Flow Machines, and Watson Beat, all of which have attempted to develop intelligent algorithms for every aspect of the music-making process, including music analysis, composition, performance, sound processing, audio streaming, and even education. Often, the emergence of new tools for AI revitalizes long-standing debates over human agency in the face of intelligent machines. Among musicians and music producers, the fear is that these new intelligent assistants will take their jobs, while AI startups claim that their tools are simply meant to free the artists from the mundane and repetitive tasks of musical labor. The Spotify Creator Technology Research Lab claims, "We've spent a lot of time asking ourselves what other technologies could enable artists to create in different, exciting ways."[3] The developers of Watson Beat write, "When you've got an exciting story to tell about technology, what better soundtrack than an original composition created with help from a cognitive machine?"[4]

In the late 2010s, these questions targeted AI mastering "assistants"—software plug-ins and online services that put the finishing touches on a final mix and prepare it for online distribution. While much of the history of mixing and mastering has emphasized the highly specialized skills and critical listening capacities of human audio engineers, automated tools are claimed to finalize a mix at a level comparable to the work of industry professionals.

Music tech company iZotope describes its vision for the future of what it calls "assistive audio technology": "Since the 2016 release of Track Assistant in Neutron, iZotope has been developing assistive audio technology designed to remove the guesswork from audio production and make it more efficient. One of our major goals as a company is to find solutions to eliminate time-consuming audio production tasks for our users so they can focus on their creative visions."[5] As in most debates about computational creativity, concerns among music producers stem from the perceived ability of the software to replace the labor of the human mastering engineer, thus dehumanizing the creative process. While I agree that this is a crucial question—one that hints at broader ethical issues concerning the relationship between humans and machines—I'm also interested in the more implicit question about agency at the root of all forms of computational creativity. That is, how might the visual aesthetic and technical affordances of AI mastering assistant software shape the fundamental experience of listening, from the perspective of both humans and machines? Further, how might these new practices of deep learning/listening by humans and machines mirror similar shifts in the labor practices of musicians and producers?

Though not always rooted in the science and techniques of AI, the history of the music products industry has seen a variety of "music assistants" come and go. From early drum machines to virtual drum emulators, groove algorithms, and randomization tools, experimental composers and producers in the commercial music industry alike have been interested in the idea of automated studio "assistants" since the early twentieth century at least. It wasn't until the twenty-first century that the industry witnessed the emergence of "intelligent" commercial software, further reflecting shifting conceptions of technology and the formation of what Alexander Weheliye calls "desiring machines" in popular music.[6] How have these tools mirrored changing understandings of what it means to be human and what constitutes music in the age of software?

While automaton musicians and automated musical devices have been around since the early thirteenth century, the drum machines of the early twentieth century established a clear lineage in form and function to many of the musical aspirations of producers to this day.[7] When composer Henry Cowell asked Léon Theremin to build an instrument that could perform polyrhythmic compositions based on the overtone series, he was looking for a machine that could do something deemed impossible by human musicians and existing instruments. Although Theremin's Rhythmicon from 1932 may not have had a lasting effect on the development of programmable drum machines, it exposed fundamental needs and desires that music producers carried with them through the development of assistive audio technology. Most important was the idea of the machine as a supplemental, yet intimate, partner to the human musician, as well as the significance of rhythmic presets as a core mechanic of the instrument. The preset provided an impetus in the commercialization of music technology, as well as a marker of big data's effect on the maximalist and accumulative nature of contemporary music production.

In 1949, Harry Chamberlin's Rhythmate introduced the idea of sampling to the domestic practice of family sing-alongs, incorporating fourteen selectable tape loops of rhythmic instruments performing various musical patterns. While the device only sold one hundred units, the idea of sample-based rhythmic presets would become increasingly popular throughout the twentieth century, especially after the commercial introduction of machines such as the Linn LM-1 sample-based drum machine in 1980 and the Akai MPC 60 sampler, drum machine, and sequencer in 1988. Wurlitzer's Sideman from 1959 was unique in that it operated in a similar manner to a music box, generating sounds through a rotating disk whose tempo could be controlled with a mechanical slider. Of note, though, is the criticism surrounding the instrument's use in professional club settings. More successful than earlier rhythm accompaniment machines, the Sideman drew heavy criticism from musicians' unions that feared the device would devalue human musicians since venues could technically pay one person for the labor of three performers. This anxiety allowed the unions to prevent the use of the machine in cocktail lounges unless the keyboardist was paid the wages of three musicians. Other early developments included the Keio Minipops (1967) and the PAiA Electronics Programmable Drum Set (1975), both of which paved the way for the increasing miniaturization of rhythm machines and eventual remediation of these tools in software. Last but not least, we can thank devices such as the Roland TR-808 (figure 7.1) and the long line of Akai samplers (figure 7.2) for establishing the enduring design layouts that survived the shift from hardware to software: the sixteen-step sequencer grid, the sixteen-pad rubber grid layout, an emphasis on portability and immediate feedback, and a standardized set of interface elements for sonic manipulation.

While drum machines and samplers exhibit a rich and diverse history of offering both generic and abstract tools for drum sequencing and editing, the

Figure 7.1 Roland TR-808 drum machine

Source: https://en.wikipedia.org/wiki/Roland_TR-808.

Figure 7.2 Akai MPC 60 music production center

Source: https://www.polynominal.com/akai-mpc60/.

desire to emulate the human feel of a drummer performing on an acoustic drum kit didn't go away in the transition from analog to digital. With the shift from hardware to software, feature creep made its way into the seemingly endless abyss of fine-tuning and parameter manipulation encouraged by DAWs. Virtual drum instrument software promised to provide both "realistic" drum sounds and "human" performances of those sounds. FXpansion's BFD ("Big Fucking Drums," 2003) was a hugely influential drum sample library that included a whopping nine GB of sampled drum sounds. Soon after the release of BFD, Toontrack joined the competition with DFH (Drumkit from Hell) Superior in 2004, which offered not only thirty GB of sounds but also the ability to control the mic distance and volume articulation for each sound. Thirteen years later, Toontrack released Superior Drummer 3, which incorporated more than 230 GB of sounds, recorded with an additional eleven separate room mics in order to allow full sonic control of not only the drum kit but the acoustic character of the studio space. FXpansion's BFD3 (2013) claims to provide "effortless production of drum tracks indistinguishable from the real thing" and incorporates what the maker calls an "augmented realism" feature which models the resonance and bleed of the toms to blend the sounds of the kit together (figure 7.3).[8] The rhetoric of effortless production, sonic fidelity, and increasingly large data sets served

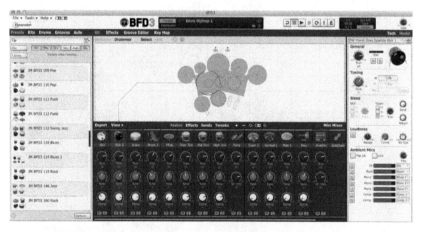

Figure 7.3   FXpansion's BFD3 drum kit virtual instrument

to foreshadow the primary marketing incentives of AI assistants just a few years later.

For those musicians and producers less interested in emulating the sound of acoustic drum kits but still very interested in "humanizing" their electronic drum sequences, the advent of what I call automated groove mechanics became a staple of DAWs in the early twenty-first century. With the release of Ableton's *Live* version 8 (2009), the software allowed producers to drag and drop preset "groove" files from their library to individual audio clips, affecting the most minute elements of the audio files' rhythmic feel (figure 7.4). In addition, the software allowed the extraction of grooves (defined by rhythmic timing variation and dynamics) from recorded or imported MIDI and audio clips, which could then be placed on any other clips in a producer's session. Similarly, Apple's *Logic Pro X* "Groove Track" (2013) aligns the rhythmic quantization of a selection of audio or MIDI tracks to that of a chosen master track. Essentially, these tools allowed the producer to "sample" and duplicate the feel of a new or existing musical idea and turn it into a sonic stamp—or

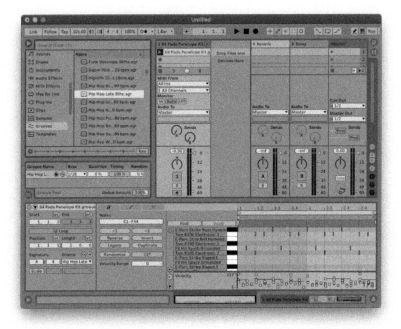

Figure 7.4 Choosing a groove from *Live*'s "Swing and Groove" library and applying it to a musical pattern (screenshot taken by author).

what musicologists Anne Danielsen and Ragnhild Brøvig-Hanssen call a "digital signature"—on other elements of the mix.[9]

This type of obsessive micro-manipulation and hyper-quantization of rhythm would pave the way for a next generation of groove tools and eventually the introduction of AI-based musical assistants. WaveDNA's *Liquid Rhythm* (2010) claims to offer a collection of rhythmic building blocks that allow you to create "up to 10 quadrillion unique rhythm patterns . . . in seconds," using its "intuitive Music Molecule Technology" to visualize your beats.[10] Accusonus's *Rhythmiq* (2019) is a self-proclaimed "A.I. Beat Assistant" that performs beats "in real-time like when you are jamming with a human band-mate" and provides "unlimited beat variations with the turn of a single knob," allowing the producer to "instantly create fills and build-ups" (figure 7.5).[11] Like the virtual drum instrument software, these tools employ a contradictory marketing rhetoric of speed, precision, and effortlessness, while

Figure 7.5 Accusonus's *Rhythmiq* AI "Beat Assistant"

Source: https://pluginfox.co/products/accusonus-rhythmiq?variant=
30274232680580&currency=USD&utm_medium=
product_sync&utm_source=google&utm_content=
sag_organic&utm_campaign=sag_organic&utm_campaign=
gs-2020-04-03&utm_source=google&utm_medium=smart_campaign.

also invoking the values and virtues of musical labor performed by human musicians.

The late 2010s witnessed a revival in research on the creative application of AI tools, especially in the world of music production. The suite of plug-ins available in Google's *Magenta Studio* allows producers to incorporate machine learning techniques for music generation inside DAWs.[12] "Continue" uses recurrent neural networks (RNN) to generate extensions and variations on melodies or drumbeats based on an existing musical pattern. "Generate" produces new musical ideas using a variational autoencoder (VAE) that has learned a summarized representation of musical qualities from millions of existing melodies and rhythms. "Interpolate" also uses a VAE but takes two musical ideas (drumbeats or melodies) as inputs to generate a set of new musical clips that combine the qualities of the input clips. "Groove" achieves a similar effect to that of the "humanize" and "groove" presets previously discussed but accomplishes it by analyzing the rhythmic nuances of human drummer performances and using those data to predict how unquantized the rhythm should be in order to emulate a specific musical style. "Drumify" creates a drum groove based on the rhythm of a second musical input (figure 7.6).

Sony's *Flow Machines* is described as an "AI assisted music composing system" that can generate melodies, chords, and basslines for use within a DAW-style workflow. The vision behind *Flow Machines* is "Augumented [sic] Creativity," as the project website claims: "Although it is often said that AI might replace human, we believe that technology should be human centered designed. We will keep on researching and developing to let technology augument [sic] human creativity even more."[13] Echoing the marketing

Figure 7.6 Google's *Magenta Studio* machine learning music generation plug-ins
Source: https://magenta.tensorflow.org/studio/.

rhetoric of augmentation behind the virtual drum instrument software previously discussed, Sony is aware of the perceived threat to human labor that AI introduces. The language of "augmentation" redirects conceptions of AI as enhancing, rather than controlling, the producer's creative workflow.

Other tools are a bit more niche. *MuseNet*, created by Elon Musk's research organization, OpenAI, generates music by combining different styles of music. The interface is simple: two drop-down menus allow the user to compose music in the style of one composer starting with the musical ideas of another composer. For example, "Compose in the style of Lady Gaga starting with Beethoven's Für Elise" (figure 7.7). The developers state up front that "MuseNet was not explicitly programmed with our understanding of music, but instead discovered patterns of harmony, rhythm, and style by learning to predict the next token in hundreds of thousands of MIDI files."[14] The project website even includes a cluster visualization that claims to display the relationships between composers throughout the history of Western music (squarely placing Katy Perry on the opposite end of the spectrum from Frédéric Chopin). While most AI tools claim—in their marketing rhetoric, at least—to value musical genre and cultural aesthetics in their deep learning algorithms, *MuseNet*'s overt emphasis on data over music theory reveals a fundamentally scientific, rationalist approach to composition that forms the core of most AI music assistants.

By the end of the first two decades of the twenty-first century, AI-assisted compositional tools existed mostly in research labs on the fringes

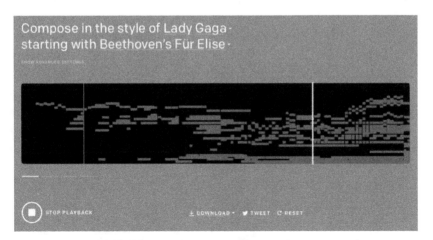

Figure 7.7 OpenAI's *MuseNet* music composition generator
Source: https://openai.com/blog/musenet/.

of actual musical practice. At the same time, algorithmic creativity experienced a renaissance in the world of automated mixing and AI-assisted mastering tools. While tools such as *Magenta*, *Flow Machines*, and *MuseNet* claim to provide inspiration and ideas for the compositional process, online mastering services such as LANDR (2014), CloudBounce (2016), and eMastered allow users to upload their audio files and have them automatically mastered for distribution to Spotify, Apple Music, and other digital music streaming services. LANDR promises "professional audio mastering with instant results . . . without paying studio rates or learning complex plug-ins." Like those of most AI music tools, its website foregrounds the computational power of the tool over musical value, emphasizing that "With years of research, 19 million mastered tracks and over 1 million hours of music, Synapse is the most sophisticated AI-powered mastering engine yet."[15]

For anyone with even a cursory understanding of the music mixing and mastering process, instant mastering tools such as LANDR and CloudBounce might seem counterintuitive. Mixing and mastering are typically positioned as specialized skill sets that require years of audio engineering training and the development of critical listening capacities (or what mastering engineers refer to as "big ears"). Mastering engineers fine-tune the nuances of a recorded mix, often starting with an obsessive calibration of their listening environment through a careful placement of studio monitors. From there, they carefully sculpt the frequency spectrum to focus the bass and make the high end crisper, erase any noise hidden in the recesses of the mix, boost the loudness of the recording to make it competitive in the market, and copyedit the digital file's metadata to ensure proper playback through all possible audio systems. Further, their training involves an extensive historical understanding of how changes in sound playback devices constantly shape the tools and techniques of music production and consumption. For example, with the advent of the MP3 format in the mid-1990s, followed by an increasing shift to mobile music listening among consumers, mastering engineers had to figure out a way to optimize their mastering workflow to ensure the audibility of as much as possible of the frequency content of the compressed audio file even when heard through small plastic earbud headphones or cheap laptop speakers. Indeed, one of the best things one can do to train oneself in audio mastering is to practice the type of deep learning and deep listening that AI mastering tools employ: listen to as much music as possible from a diverse range of styles and traditions, keep up to date with the latest tools and technologies of audio production and playback, and develop a vast comparative and analytical mindset toward musical recordings.

As these examples attest, the question for the future of music production isn't about the looming automation crisis (contrary to popular belief). That's already here. The question is, how is AI affecting the nature of work in recording, mixing, and mastering studios? Assistive audio technologies are marketed for their speed and maximal productivity, which reflects the general push among employers for greater efficiency, speed, and productivity in their employees. Amper's *Score* allows the user to "create your own original music in seconds."[16] CloudBounce goes so far as to anticipate user concerns about humans losing their labor to AI. Its tagline, "Built by musicians, for musicians," reminds consumers that they're the good guys: "Automatic processing can never replace the final touch of a professional mastering sound engineer. However, with CloudBounce, you can get almost there—cost effectively."[17] Yet on the same web page, it follows the trend of other AI tools by claiming that "crystal clear and airy sound quality" can be achieved "without the need for training or tools," a major slight to professional mastering engineers who have dedicated years of training to the practice of critical listening.

Assistive audio technology restructures the balance of agency within the studio, as software takes on a managerial role and the engineer becomes the assistant. This trend mirrors the labor practices of corporate warehouses around the world. As journalist Josh Dzieza describes in his article "How Hard Will the Robots Make Us Work?" the hyper-efficiency of intelligent machines has created maximalist labor conditions for workers across industries. At Amazon, "Almost every aspect of management at the company's warehouses is directed by software, from when people work to how fast they work to when they get fired for falling behind. Every worker has a 'rate,' a certain number of items they have to process per hour, and if they fail to meet it, they can be automatically fired."[18] These types of warehouse labor practices are becoming increasingly common, built from Amazon's model in which workers become the archetypal robots, literally standing in cages while deshelving items brought to them by actual robots. Automated management systems function by simultaneously distributing labor across multimedia venues (warehouse, call center) while using surveillance systems as a sort of panopticon that encourages maximal productivity. The fear of AI stealing human jobs—most often peddled by Silicon Valley tech giants—is not only wrong, according to Dzieza; it's a distraction used to prevent the horrific labor practices of the tech industry from being revealed.

Outside of material labor practices, the loose intellectual property considerations behind deep learning algorithms have significant implications for creative ownership and artist compensation. In many ways, the software presets curated from the machine's deep listening to millions of commercial recordings function as a type of next-generation sampling, in which the "intelligent assistant" samples actual human labor without compensating the original musician. Music producer and journalist Dani Deahl notes that "the law doesn't account for AI's unique abilities, like its potential to work endlessly and mimic the sound of a specific artist." In this context, "How can anyone prove an algorithm was trained on the song or artist it allegedly infringes on? . . . Current US copyright law doesn't differentiate between humans and non-humans."[19]

A core premise of this chapter has been the idea that AI's deep learning algorithms function as a type of machine-oriented deep listening. In this context, the actual software interface of automated mastering tools doubles as the social and cultural interface between machine and human aesthetics in the practice of listening. Analyzing the design and layout of tools such as iZotope's comprehensive mastering software *Ozone*, we observe a heightened interdependence between human and machine agents, as well as an increasingly diagnostic mode of listening in which engineers focus their attention on repairing flaws in the mix. Aden Evens describes this type of obsessive, hyper-rational listening mode in the practice of digital music production: "At a remove from the sound, standing over it, the electronic musician reflects on the sound, has the opportunity and the distance to hear every detail. Digital music tools allow and encourage an unending editing process, exposing every aspect of the sound to the music maker and offering a focus on arbitrarily small detail and arbitrarily large structure. . . . Hours a day in front of the keyboard and mouse are spent *playing* a piece."[20] In the context of iZotope's high-resolution displays and the mastering engineer's fine-tuned attention to detail, diagnostic listening functions as a type of hyper-"atomized" listening, following Theodor Adorno's critique of mass-media technologies in the early twentieth century.[21]

Fundamentally, iZotope's mixing and mastering software programs are designed as monitoring systems, just as the primary goal of audio engineers and producers is to monitor, listen, and attune themselves to the current status of the music. With the emergence of assistive audio, intelligent machines, and smart media, we might ask, what are we really listening to? Technical and scientific details of digital audio, such as the lack of presence in the low-midrange frequency spectrum? Aesthetic considerations such as the effect of the compressor on the overall dynamic range of the mix? Social and cultural issues concerning the labor of human agents in the past hundred years of musical recording, engineering, and producing? In his chapter on "Listening" from *Noise,* social theorist Jacques Attali views music as a herald of what society will eventually become, having always served as a mirror of the sociopolitical reality of its time: "Listening to music is listening to all noise, realizing that its appropriation and control is a reflection of power, that it is essentially political."[22] Mastering engineers in the 2010s became soldiers in what many referred to as the "Loudness Wars," appropriating noise in order to make music compete in the neoliberal marketplace.[23] What was being fought for?

The core mechanic of iZotope's assistive audio technology is the company's extensive corpus of musical mixing and mastering techniques analyzed through the deep learning of tens of thousands of musical recordings. iZotope CTO Jonathan Bailey describes some examples of how deep learning is used in the software:

> One example from Neutron, our intelligent channel strip, uses deep learning to identify ("classify") which instrument is represented by the audio in any given track in your music session, and based on that categorisation, and some additional acoustical traits we analyse within the audio, we make a recommendation for which dynamics, EQ and/or exciter settings to apply to prepare that track for your mix. We're now using deep learning to not only analyse audio content but also process it. In our recent release of RX 7, the Music Rebalance feature uses deep learning to "unmix" a musical mixture into individual stems that can be rebalanced or otherwise processed separately. We're exploring how deep learning might be used to synthesise content in the future.[24]

The practice of curating musical content through techniques of "big data" gathering is not unique to iZotope. Techniques of music information retrieval (MIR) have used deep learning algorithms to develop music recommender systems in streaming platforms, separate tracks in a mix for DJ software, automatically transcribe musical recordings, and generate original music ideas.[25] This has radically reshaped everyday practices of musical listening, introducing us to what technologist Marinos Koutsomichalis terms the era of "big music," referring not only to huge collections of music files but also to the increasing variety of interfaces through which audio content is accessed. "Given audiences' need to orientate themselves amidst such an abundance of ever-available digitized music," Koutsomichalis writes, "listening has eventually become entwined with music information retrieval and with social media networking."[26]

One consequence of big data is the exponential increase in scale and scope of how digital subjects view their social and psychological status, as well as their interrelationships. As a result, many of the interfaces built on big data sets foster highly specialized modes of self-surveillance and self-regulation. Similar to the way social-media feeds function as what some might call "black mirrors" to society, reflecting to us the darker aspects of interpersonal life, iZotope's GUIs encourage a diagnostic listening mode in which the sonic signal is treated like a neoliberal subject, blurring the distinction between subject and object, producer and consumer, art and science. In his famous critique of the contemporary culture of art and design, art critic Hal Foster asks a core question: to what extent has "the constructed subject" of postmodernism become "the designed subject" of consumerism? Twenty-first-century "Design" is, for Foster, defined by the conflation of marketing and design, commerce and aesthetics: "when the aesthetic and the utilitarian are not only conflated but all but subsumed in the commercial, and everything— not only architectural projects and art exhibitions but everything from jeans to genes—seems to be regarded as so much design."[27] In this context, iZotope's extensive corpus of tens of thousands of musical recordings analyzed through machine learning can be viewed as the ultimate integration of art and commerce, production and consumption. Deep learning's reliance on big data perpetuates Foster's "design subject" by masking algorithmic biases that create normalizing hierarchies of genre and taste constantly pushed by the music industry itself. We might think of how the vast presets modeled from millions of musical recordings function as what technology critic Sara Watson calls "data doppelgängers," confusing the ontology of what constitutes the musical work and who owns the musical labor.[28]

On one hand, the diagnostic design of iZotope software reifies the rational functionalism and objective listening stance often peddled by mastering engineers and reflected in the design of many of the software applications discussed throughout this book. The multi-windowed monitoring workspace of *Ozone* 9, for example, privileges visual displays that echo the diagnostic GUIs of medical interfaces. In contrast to the responsive, clean, and flat design aesthetics that became popularized with the emergence of smartphones in the late 2000s, medical interface design relies instead on the sheer functionality of Web 2.0 design principles. Skeuomorphs are encouraged over flat design; the use of text is encouraged, rather than signifying meaning solely through icons; static pages are privileged over the dynamic effect of parallax scrolling; traditional keyboard controls such as up/down arrows and the numerical keypad are perceived to alleviate errors that may arise from touchscreen or hardware "spinner" knobs; and stylized typefaces are discouraged.[29]

While contemporary digital medical interfaces use a combination of sound and image as diagnostic markers for the personal health of an individual, sound functioned as a primary diagnostic marker in the predigital era. Historian Karin Bijsterveld describes how workers used sound to diagnose the effective operation of machines.[30] As machines became less noisy over time, operators began to use visual information more often. Now the visual display has become, arguably, the primary diagnostic marker in evaluating the "status" of the sonic signal in DAWs and VSTs. Record production expert Alan Williams writes:

> The display is guide, tool, advisor, and educator. As the vertical line travels across the screen, it tells everyone present what to listen for, and when to listen for it. . . . Such design makes audio manipulation a more fluid and intuitive process, but the graphic display also has the perhaps unintended consequence of educating any observer in some of the finer points of recording science. Difficult to grasp (and even more difficult to communicate) concepts such as multi-band equalization and compression become more readily understood by anyone watching the screen. . . . Perhaps the reason this public display has such a powerful effect is that the mechanically generated illustration provides proof of quantifiable performance flaws. The subjective hunch becomes a scientifically verified "fact."[31]

This passage highlights a few important factors behind the social and cultural logic of music software, especially in the context of mastering tools. First is the idea of the AI assistant as more than just an assistant but a teacher and guide that shapes the direction of the producers' creative workflow. Second, this pedagogical function has the consequence of democratizing the techniques of music production, which is echoed in the marketing rhetoric of assistive audio tools themselves. Third, the visual display provides "proof of quantifiable performance flaws" and "scientifically verified 'fact.'"

Pitch correction software is the go-to case study here, as the visual display of the interface encourages an obsessive scientific precision that makes both producer and performer hyper-aware of performance "flaws" (the software could be said to create the flaws, in this same way). Of course, this aspect of music software design extends outside of musical performance and is even more apparent in the diagnostic interfaces of assistive audio tools.

Ultimately, the self-awareness and self-surveillance that visual displays encourage reflect what Btihaj Ajana calls the "biopolitics of the quantified self," whereby the body is made amenable through self-monitoring technologies to management and monitoring techniques that often echo the ethos of neoliberalism.[32] Again, we might think of the autotune effect in which singers learn to hear every imperfection in their voice after being exposed to the obsessive visual manipulation of their pitch, formant, rhythm, sibilance, and more through pitch correction software such as Antares's *Autotune Pro* and Celemony's *Melodyne*.[33] This phenomenon extends to casual listeners experiencing pop music records from the mid-twentieth century as being "out of tune" to their twenty-first century ears that have learned to anticipate pitch correction in recordings as the norm.

The biopolitics of assistive audio are most readily apparent in GUIs replete with visual monitoring displays directly reminiscent of medical interfaces, most often employed for the purposes of sonic control, containment, and repair. In iZotope software, the perception of sonic control is made apparent through big, colorful meters that call to mind the fantasies of control perpetuated in sci-fi console design. The apex of diagnostic design can be found in iZotope's *Insight* 2, an "intelligent metering" software that analyzes loudness, intelligibility, spectral balance, and more "with surgical precision" and "in stunning 3D."[34] Visual displays mark the health of the audio signal, all of which strive for normalization: red numbers denote higher-than-normal loudness levels, green meters let you know that a given element of the mix is "intelligible," bright white areas of the spectrogram and sound field graph mark density of frequencies and stereo spread, respectively (figure 7.8). ▶

In *Ozone*, each effect in the mastering chain incorporates responsive, high-resolution meters with the option to choose from multiple graph styles and visual perspectives. The *Ozone* "Imager" visualizes the stereo space of the mix as small blue dots rapidly moving across a 180-degree fanlike graph. "Dynamic EQ" incorporates colorful nodes across the frequency spectrum, with white lines floating up and down in response to the real-time frequency content in the mix. As listeners view these dynamic, real-time visualizations

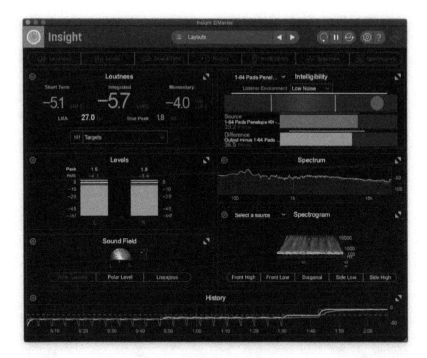

Figure 7.8 iZotope's *Insight* 2 "intelligent metering" software (screenshot taken by author).

that move and breathe in line with the sonic mix, they're given the impression of diagnosing not only the sonic health of a musical recording but the actual health of a living, breathing entity (figures 7.9, 7.10). These biopolitics are reflected in the titles of *Ozone*'s presets, many of which convey a language of repair, emphasizing a goal even more obsessive in the mastering process than in the recording or mixing phases. "Clean Limiting," "Clean Low End," "Dynamic EQ Clean Up," "Control Dynamics," "Control Mud," "Remove Mud," and "Mud Cleanse," among others, all amplify a need to foster a perfectly controlled acoustic space.[35] ⊙

As much as plug-ins such as *Insight* and *Ozone* reflect a diagnostic design and biopolitical listening mode, iZotope software also incorporates design elements that encourage active, comparative musical analysis and deep listening based on the software's extensive corpus of deep learning examples. The "Listen" button within the "Low End Focus" module (figure 7.11) makes audible the sonic difference between the processed and unprocessed signal while clarifying the low-end frequency spectrum. The "Match EQ" module

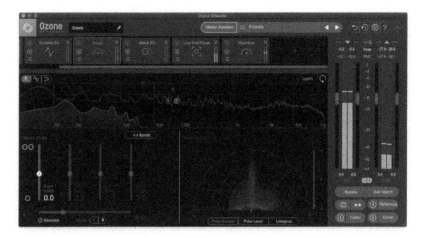

Figure 7.9 iZotope's *Ozone* "Imager" (screenshot taken by author).

Figure 7.10 iZotope's *Ozone* "Dynamic EQ" (screenshot taken by author).

(figure 7.12) takes a sonic snapshot of the graphic equalizer curve in an uploaded reference track and applies it to the current mix. The "Reference" section of the software provides a visual comparison and an A/B switch to compare the user's mix with multiple uploaded audio files (figure 7.13). Yes, these functions reify the biopolitics previously mentioned: pitting recordings against each other to better compete on the global marketplace, capturing the micro-details of a mix in order to diagnose problem frequencies, making

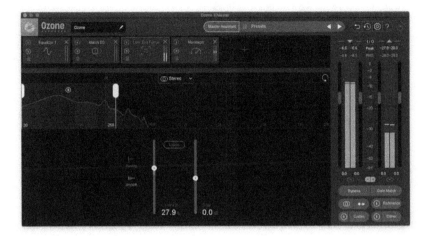

Figure 7.11 "Listen" button in the "Low End Focus" module (screenshot taken by author).

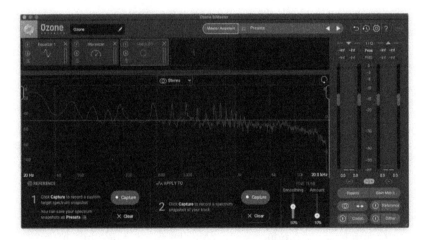

Figure 7.12 "Match EQ" module (screenshot taken by author).

audible the practice of sonic repair as personal improvement and "progress." At the same time, their pedagogical function is useful to students and other users who may not have access to the expertise and tools of a professional mastering engineer. Despite the potential of presets to create carbon-copy mixes and discourage more active techniques of deep listening, they also offer a starting point for beginners to learn how to add attack, focus, warmth,

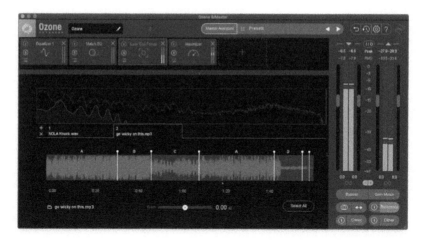

Figure 7.13 "Reference" section with two reference tracks uploaded (screenshot taken by author).

and brightness to their mixes, elements of a twenty-first-century commercial music recording aesthetic that consumers have come to expect consciously and subconsciously in their listening practices.

With seemingly infinite variation and nuance, you'd think assistive audio technology could live up to the dream of AI and not only emulate human capacity but extend and enhance it. Often, though, critics claim that the results sound too much like cookie-cutter presets, watered-down copies of technical affordances grossly translated from hardware to software. In a scathing critique of LANDR's partnership with SoundCloud, arts and technology writer Sam Machkovech begins by evoking the perennial robot-driven automation crisis: "First, the robots came for our factory jobs, then for our fast-food jobs. Now, they're aiming their laser-guided sights at . . . music-recording engineers?" After referring to AI mastering services as "snake oil" and quoting a dozen musicians and producers who consider them to be "shit," his conclusion states simply that "SoundCloud and LANDR users can expect to get what they pay for."[36]

In some ways, the skepticism toward AI mastering services among actual users is ironic, since so many mastering engineers see their own roles and responsibilities in a similar vein to the AI assistants: impartial, objective, guided by the countless amounts of "data" trained into their ears over years of experience. Pedagogical texts often describe the art *and* the science of mastering, but many engineers are quick to prioritize the objective, scientific aspects of what they do. Mastering engineer Rob Stewart acknowledges mastering as art and science but claims it to be "Science first. To be effective, the mastering engineer needs to understand and consider psycho-acoustics (how our brains perceive and process sound), along with many other technical aspects of sound (e.g. the physics of sound) and audio (e.g. electronics, signal flow, etc.)."[37] Audio engineer Bobby Owsinski writes about the value of professional mastering engineers in the age of assistive audio: "You're paying for the ears and experience, which can't be found in a plugin with a preset. You're paying for clean and precise signal path and playback environment, which is very unlikely to be duplicated at home. You're paying for an external trusted opinion and taste that a DIY job can never provide."[38]

Despite the pervasiveness of objectivity, technicality, and precision as core principles and values in the mastering process, language about the dark art and hidden mystery of mastering is just as common. As audio researchers Brecht De Man and Joshua D. Reiss note, the very books that stress the scientific principles of digital audio—outlining objective parameters, sonic characteristics, device settings, and so on—stress that mixing is a "highly non-linear," unpredictable business, devoid of "hard and fast rules," "magic settings" or "one-size-fits-all equaliser presets."[39] In a *Pitchfork* feature titled "The Dark Art of Music Mastering," journalist Jordan Kisner writes, "If rock stars are the sex gods of music, mastering engineers are its druids, the ones who work methodically and meticulously, and to whom people come for mystical wisdom and blessing."[40] M.C. Schmidt of the experimental electronic duo Matmos claims mastering to be "such a dark art," while his partner, Drew Daniel, describes it as "this mysterious process that a lot of musicians don't understand, including us."[41] In a review of *Ozone* 9, journalist Alex Holmes sums up the "Master Rebalance" module in one word: "witchcraft."[42]

This alternative, "unscientific" approach to musical listening might hold the key to where assistive audio technology needs to go in the future. On a concrete level, we might follow De Man and Reiss's advice for AI engineers to build more "high-level" information such as instrument information, studio recording conditions, playback conditions, genre, target sound, and more, into deep learning algorithms. As they argue, incorporating this type of knowledge into assistive audio algorithms "shifts the potential of automatic mixing systems from corrective tools that help obtain a single, allegedly ideal mix, to providing the end user with countless possibilities and intuitive parameters to achieve them . . . end users would have a sonic equivalent of Instagram at their fingertips."[43] Audio engineer Bernie Grundman talks about the potential for human-computer empathy in the mastering process, as engineers embrace "this willingness to enter into another person's world, and get to know it and actually help that person express what he is trying to express, only better."[44] On one hand, machine listening and "big music" data mining seem to reflect a type of "ubiquitous listening" in which listeners become subject to the neoliberal biopolitics of big data surveillance and free labor, providing hours of listening time just to churn out a preset track that reifies the original data set.[45] At the same time, *Ozone* reflects a nuanced approach to listening that focuses obsessively on the micro-details, echoing something akin to a machine-oriented deep listening, following Oliveros.[46] This empathetic mode of listening may hold the key for the next generation of AI, in music and digital culture more broadly.

Conclusion

Invisible Futures

Nobody expected the dumpster fire of 2020. It started with a global pandemic as coronavirus shuttered most of the world. Suddenly, many of us were forced to navigate an emerging social, cultural, economic, and political crisis entirely through our smartphone screens and computer monitors. Those who could had to work from home, adopting virtual productivity apps and video conferencing software, while large tech corporations cashed in on what many referred to as "the new normal." Facebook, Google, and pretty much every other company in Silicon Valley reassured us that they were there for us in those trying times, and artists quickly took them up on that offer, posting thousands of hours of virtual performances and live-streamed events amid the closing of in-person concert venues. In schools, K–12 music directors suddenly became video editors, splicing together dozens of hours of student performance footage in the trendiest mode of performance in 2020: the "virtual choir." The image of a video grid matrix populated with a diverse and enthusiastic community of vocalists all singing in perfect postproduced harmony came to exemplify sonically and visually the core zeitgeist of what journalist and activist Naomi Klein called "coronavirus capitalism," in which large companies absorbed massive amounts of federal aid and a boost in product demand while individuals and small businesses struggled.[1] Scrolling through the endless ephemerality of Facebook feeds, the sea of virtual choirs reminded me of a quote echoed by everyone from Donald Trump to Wonder Woman to the United Nations: "We're all in this together."[2]

Both in the fabricated community of the virtual choir and in the fabricated community of a global society in quarantine, each passing day of 2020 reminded us of the cracks continually being created by Covid-19: increasing rifts in the ability of different socioeconomic classes to navigate the pandemic, increasing political rifts caused by wearing (or not wearing) face masks, increasing discussions about what constituted "real" or "virtual" communication in the age of social distancing. Throughout most of the year,

Push. Mike D'Errico, Oxford University Press. © Oxford University Press 2022.
DOI: 10.1093/oso/9780190943301.003.0009

many casual viewers believed virtual choirs were happening in real time, seeming to confirm the idea that it's possible to engage in meaningful social, cultural, and intellectual interactions completely online. Like coronavirus capitalism itself, though, these videos were carefully edited, produced, and marketed to boost the viability and cultural life of the institution being represented, as a demonstration to the world that—through an exceptional social and technological infrastructure—this school, business, nonprofit, and so on, was uniquely able to outlive the virus.

Most of the "social collaboration" tools for virtual music making that emerged at this time—the *Acapella* mobile app (figure C.1), *JamKazam* (figure C.2)—followed the virtual choir model, allowing users to create multitrack video recordings by sharing audio and video files through cloud-based

Figure C.1 *Acapella* (2015) "virtual choir"–style app (screenshot taken by author).

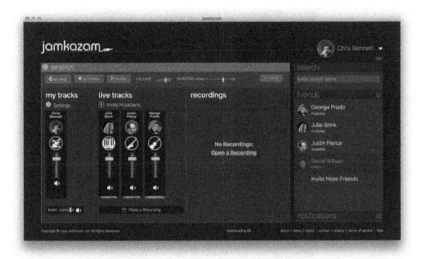

Figure C.2 *JamKazam* (2020) app "In Session" page

Source: JamKazam, "Screenshots," accessed August 31, 2020, https://www.jamkazam.com/corp/scre enshots.

file repositories. ⊚ Other tools claimed to offer actual "real-time" jamming and musical collaboration capabilities. In the spirit of *Ocarina*, music software developer and musician Tim Exile created *Endlesss*, a "lightning-fast collaborative music-making app" that allowed users to "make beats, improvise music and jam live with musicians, beat-makers and DJs all over the world" (figure C.3).[3] The app tagline promised "boundless music creation. *Alone or together*," calling to mind Sherry Turkle's neologism to describe a wired social-media generation whose strongest sense of togetherness came from experiences with which they engaged behind small screens. ⊚

As most of the United States began feeling the cabin fever of quarantine, posting on social media shifted from a leisure activity to a hub for one's social, economic, and political identity, and consumer tech seemed to offer the promise of social engagement outside of one's living room. Infamous Twitter troll Donald Trump used this moment as a political opportunity, fanning the flames of red-state discontent by blaming the coronavirus on China, decrying attempts by state governments to quell the virus in the absence of federal help, and offering his expert medical advice to avoid the virus, which included drinking toxic cleaning supplies. This was all just a prelude to what came next. Just as people's Twitter feeds started to shift away from #coronavirus, the start of summer 2020 in America witnessed a string of police

Figure C.3 *Endlesss* (2020) "Drums" mode (screenshot taken by author).

violence that culminated in the murders of Breonna Taylor in Louisville and George Floyd in Minneapolis. In the months following these events, protests and riots against racism and white supremacy engulfed much of the Western world, and by July, federal military contractors were being sent to left-leaning cities like Portland, Oregon, to shut them down violently.

It's no coincidence that video game consoles and VR headsets were sold out for most of the year. Video game website *Polygon* claimed that "Summer 2020 will be all about escapism."[4] I bought into it. With my daughter's day care closed, we spent many hours playing *Animal Crossing: New Horizons* (*ACNH*) on the Nintendo Switch, building our dream island, carefully curating everything from the physical architecture to the natural landscape and social community (going so far as to kick out neighbors we didn't like). We weren't alone. In the month of March, the game sold 5 million copies, breaking the record for the highest-selling game for any platform in a single month. Countless news outlets referred to *ACNH* as "the game for the coronavirus moment," "exactly what we need right now," "the perfect game for self-isolating," and "the perfect pandemic pastime."[5]

Meanwhile, Gita Jackson of *Vice* decried the unfortunate direction in which she saw the game and its players moving, "away from peacefulness, into the mad dash of a collector's mindset." As she claims, "the lesson that *Animal Crossing*, as a series, has tried to impart is that it is alright to stop rushing. It's okay to take a seat on a park bench, to be with yourself for a moment, to just breathe."[6] In many ways, *ACNH* mirrored the shift so many people were making from an analog to a digital world. As much as some saw the pandemic as a chance to rethink and retool the status quo, decolonize virtual classrooms, and make more accessible the tools of digital creativity, the emerging virtual spaces of the 2020 quarantine simultaneously intensified and reified some of the most problematic aspects of "real" and face-to-face (F2F) spaces—racism, xenophobia, homophobia, class struggle, all reflective of a late-stage capitalism at max capacity.

The VR boom of the 2010s was no different. Touted as the "first all-in-one gaming system for virtual reality," the Oculus Quest, released in 2019, promised a fully wireless and PC-free, immersive, escapist media fantasy: "No Wires. A World of Experiences. Unlike Anything You've Ever Seen."[7] Sold at a comparable price to most video game consoles, and removing the requirement of also having a high-performance PC to use the headset, the Quest made high-quality VR financially accessible to a much wider range of users. My first experience with the Quest felt surreal. The console's dashboard placed me in a beautiful glass house overlooking a vast mountain and a night sky full of stars. Other well-lit homes were visible just beyond my purview, giving me the impression that I wasn't alone. In fact, some of the most popular "games" for the Quest— *VRChat* and *Rec Room*—provided virtual spaces to hang out with other users in real time, alone/together.

Virtual tourism was a big sell for VR in the 2010s. With 360-degree YouTube videos, you could step into the shoes of climbers on Mount Everest, astronauts exploring outer space, scuba divers swimming with sharks, all from the comfort of your own headset. Or, in the less picturesque social-media uploads, you might find yourself observing some peaceful protesters getting tear-gassed by military contractors. As Ruha Benjamin notes, "One of the problems with VR is that it can present another opportunity for 'poverty porn' and cultural tourism that reinforces current power dynamics between those who do the seeing and those who are watched."[8] Despite the perceived strides in accessibility that next-gen VR seemed to promise, you couldn't find a Quest for most of 2020. During quarantine, this type of virtual travel felt like a privilege, echoing debates about who does or doesn't get to escape the global pandemic physically, emotionally, or economically.

Music-making apps on the Quest borrowed these core mechanics, prioritizing social connection, immersion, and intimacy, as well as pedagogical value. *Tribe XR* teaches you how to DJ by placing you in front of a fully functioning set of CD DJ turntables. When you're ready to go, you can upload your own music and live-stream custom sets. You can even invite friends to join your private DJ room and host back-to-back DJ sets in which you and a friend take turns spinning tracks. ▶*Electronauts* places you inside an abstract, sci-fi music production studio, allowing you to make and manipulate sounds by interacting with colorful spheres representing percussive sounds, vials of neon fluid that act as mixing board faders, and a simple grid GUI with buttons for triggering one-shot loops. Like so many digital tools in the twenty-first century, *Electronauts'* marketing materials focus on the game's ability to provide maximalist content regardless of musical skill, allowing the user to "create music with confidence." As the developers claim, "Electronauts is powered by Survios's proprietary Music Reality Engine, which transforms song elements into interactive components that sound epic no matter how you mix them."[9] "Music Reality Engines" might be the perfect descriptor for the core mechanics of so many digital music-making tools in the first decades of the twenty-first century, designed to be instantly available, infinitely manipulatable, and intimately personal. In this reality, digital audio, too, becomes fully immersive.

If the hype surrounding VR historically centered on its ability to immerse us in rich experiences and provide social connection in virtual space, VR in 2020 instead felt like pure escapism. In the digital quarantine, virtual virtues became synonymous with virtuality, and virtuality became synonymous with virality. It seemed perfectly fitting and a sign of things to come when social-media giant Facebook purchased Oculus for $2 billion in 2014. By 2020, VR represented at once the culmination and contradiction of the technological solutionism, techno-optimism, and digital maximalism of the first two decades of the twenty-first century. Yes, new interfaces for social and cultural expression allowed for increased virtual connectivity and creative problem-solving in the workplace, but—as is the nature of all interfaces— they also obfuscated structures of power, privilege, and labor behind novel mechanics, fleeting hashtags, and escapist content. This defined the invisible future of VR in 2020.

Is this not the ideological nature of all interfaces, though? Wendy Chun claims that "interfaces have become functional analogs to ideology and its critique," in that they "seem to concretize our relation to invisible (or barely visible) 'sources' and substructures."[10] Skeuomorphs, in particular, reflect what she calls the "fetishistic logic" of interfaces, in that "users know very well that their folders and desktops are not really folders and desktops, but they treat them as if they were."[11] As such, not only do interfaces use metaphors in their visual design, but the interface itself functions as a metaphor for understanding all invisible laws: "its combination of what can be seen and not seen, can be known and unknown—its separation of interface from algorithm; software from hardware—makes it a powerful metaphor for everything we believe is invisible yet generates visible effects, from genetics to the invisible hand of the market; from ideology to culture."[12] For Chun, the ideology of the interface lies in its ability to make us believe that "the world, like the computer, really comprises invisible hands and rules that we can track via their visual manifestations."[13]

This ideology defines the paradox of digital culture and HCI in the twenty-first century: the blurred line between VR as an "empathy machine" that allows us to walk in someone else's shoes and a portal into escapist fantasies; the dream of social media as a tool for immediate social connection, anytime, anywhere, versus its superficial use as a mask for the mundane lifestyles we're trying to hide from each other; digital music production as the apex of creative democratization against the backdrop of a greedy music products industry that uses maximalism and solutionism as marketing tools to encourage ubiquitous consumption. By conflating the dual role of computers as dream machines and agents of capitalist hyper-consumption, the commodity culture surrounding interface ideology constitutes what Christian Ulrik Andersen and Søren Bro Pold call "a phantasmagoria that embodies the deepest desires of its worshippers, but constantly hides its origins ... the cloud, and the cloud's metainterface, can also be seen as a disguise of capitalism that operates on the level of social and collective dreams."[14]

As much as they conceal the "real" (labor, politics, systems of value), interfaces reveal the core logic of "virtual" reality as a horizon of possibility, always looking to a future that can't be seen. An ethics of HCI emerges in our constant push and pull with software's affordances and constraints in order to shape this invisible future. At its core, this is an ethics of care and relationality, rather than the control allegories that so often get pinned on hardware. This is why Kittler was wrong when he claimed that "there is no software." The ontology of software is not simply a transcoding of physical hardware devices onto virtual representations. It's also a liminal space in which users confront a dual subjectivity, not as autonomous nor heteronomous agents but in a fluid state of what philosopher Brian Massumi calls a "transitional infinity."[15] As these users both use and are used by their tools, an ethics of HCI emerges from their inability to achieve the virtue of self-transcendence offered by technology, forcing them to constantly confront both their actual and virtual selves as finite. Even Oculus chief scientist Michael Abrash has commented on the limits of VR technology and its inability to reproduce the "real" feelings and digital intimacies that it promises: "As important as haptics potentially is for VR . . . it's embryonic right now. There's simply no existing technology or research that has the potential to produce haptic experiences on a par with the real world."[16] Understanding interfaces as ethical, relational, and transitional might allow for a reconception of what constitutes "good" gameplay, music making, and so on, in the context of software: not the user as a master of technology but rather an unmasterable user, a relational user always in between their body and its disembodiment, action and virtualization, the virtue and its realization.

In staging a self-in-relation that's simultaneously internally reflexive (individually immersive) and externally potential (virtually mediated), VR provides a space for the negotiation of ethical (digital) subjectivities. As philosopher Judith Butler writes: "If it is precisely by virtue of one's relations to others that one is opaque to oneself, and if those relations to others are the venue for one's ethical responsibility, then it may well follow that it is precisely by virtue of the subject's opacity to itself that it incurs and sustains some of its most important ethical bonds."[17]

These bonds are made possible by the recognition of the interface as a fundamentally transitional space, in which relations *between* finite identities or ways of being are set into motion. Philosopher Adriana Cavarero describes this "relational ethic of contingency" as being "founded on the *altruistic* ontology of the human existent as finite,"[18] while Massumi describes the transitional space of the interface as forcing recognition of this finitude. Following moral philosopher Tetsuro Watsuji, we may say that software, "rather than resulting in an immersion in a world where the self is lost . . . lead[s] us to a place wherein the self is found."[19] This is the major paradox of immersive media in the twentieth and twenty-first centuries.

While technology seems to offer the virtue of a limitless infinity, the interface forms the skin of the ethical subject, establishing the boundaries of an incorporeal body incapable of transcending itself. In the context of VR, virtue means not attempting to overcome the technical limitations of the material interface but rather recognizing the multiple relationships that constantly emerge at the interface between one's actual and virtual self, a contingent and liminal identity—to quote Cavarero, "a fragile and unmasterable self."[20] Butler describes this recognition as the moment of humanization: "Ethics requires us to risk ourselves precisely at moments of unknowingness, when what forms us diverges from what lies before us, when our willingness to become undone in relation to others constitutes our chance of becoming human. To be undone by another is a primary necessity, an anguish, to be sure, but also a chance—to be addressed, claimed, bound to what is not me, but also to be moved, to be prompted to act, to address myself elsewhere, and so to vacate the self-sufficient 'I' as a kind of possession."[21]

These cathartic "moments of unknowingness" define the virtue of virtuality as the unrealizable desire for transcendence of the actual ("what forms us") and the virtual ("what lies before us"). Our "chance of becoming human" comes from the willingness to give up this desire as an end in itself and to recognize our finitude as we exist in the transitional immediacy, at the interface in which the "self-sufficient 'I' " to be found nowhere in "reality"—the virtual 'I'—is constantly actualized in new formations.[22]

Virtuality is not therefore synonymous with unreality. It's a space of possibility in which the interface opens up a continuum of multiple forms of reality in which we constantly inhabit. Many of the "real" worlds we inhabit are in fact virtual realities, experimental spaces through which we attempt to achieve desired virtues. The American dream offers the possibility of autonomous, independent wealth, regardless of our position in, and relationship to, society. The stock market provides a space in which the possibility of unlimited wealth is a virtue, following media scholar Peter Horsfield's definition of VR: "a space in which we can explore the possibility of a situation in which the threats, frustrations, and limitations of our actual life do not exist."[23] Even war has become a VR, extending the technological eros of immortality through the use of remote-controlled drone attacks and video game training programs for soldiers.[24]

When these VRs fail, ethical subjects emerge, forced to recognize their finitude through a recognition of relationality. When the stock market overextends its virtual currency, as in the case of the margin call, millions lose their homes, and the self-sufficiency of the American market is questioned on a global scale. The news of war casualties provides us with a whiff of mortality, calling into question the perceived invincibility of the modern war machine. The failure to achieve the virtues of virtuality remind us that the virtuous self is a fundamentally relational subject.

Interfaces provide a platform to confront these liminal relations. Following video game theorist Jane McGonigal, virtual environments push the individual to "work at the very limits of their ability" in a state of infinite self-transition.[25] Here users reclaim the products of their labor as their own, distinct from "real world" work, in which a fear of failure often prevents them from taking these risks. Simultaneously, virtual music performance forces individuals to recognize the limits of their corporeality through a confrontation with the interface. Anthropologist Tom Boellstorff calls this "creationist capitalism," in which labor is understood in terms of creativity, so production is understood as creation.[26] In this context, digital virtuosity involves the recognition of relationality at the expense of the self-sufficient "I" to be found nowhere in reality. Virtuality involves the virtuous navigation of virtues *as* a new reality; the space in between actual and virtual, in between the arrow and its target.

The year 2020 marked a convergence of various interfaces developed throughout the first decades of the twenty-first century: social media, mobile music production, next-gen VR, and online video gaming, to name a few. As much as creative production during the pandemic gave musicians a sense of togetherness in quarantine, and as much as the virtual spaces of games and VR seemed to promise a simultaneous ability to escape from each other, the evolved ecology of digital media in 2020 had the consequence of increasingly tethering users to the endlessly scrolling media feeds and grinding labor of online, collaborative "play." Ultimately, Covid-19 reintroduced for many musicians and creative producers the core issues of this book: how software shapes us and how we use software to shape our world, how design mirrors its social context, and how our confrontations and negotiations at the interface between our inherited past and horizons of possibility define the aesthetics and politics of what comes next.

In this book, I've considered how music making and performance have been influenced by the design of music production software. I've discussed how interface design reflects broader cultural aesthetics, examined how the affordances and constraints of software are materialized in live performance with technology and everyday embodied interactions with hardware, and explored how emergent aesthetics of sound in interactive media might suggest a more dynamic relationship between the technical aspects of design and the aesthetics of user experience. Across the chapters, I've defined procedural listening as a dynamic, two-way process of HCI in which technology users focus on the structures at work in the system with which they are engaging, rather than the content created.

Having established a shared relationship between the design and use of software interfaces, we can imagine procedural listening in a new light. As chapter 2 made clear, integrating new tools into digital media production practices isn't simply about reconfiguring or remediating existing technologies and cultural aesthetics. Interfaces are more than just the material metaphors and GUIs that define our practical interactions with computers; they're a defining force in the emergence of a broader collective social and cultural consciousness in the first decades of the twenty-first century. Interfaces teach us how to relate to technology, culture, and society; interfaces shape, and are shaped by, cultural forms in an interactive way that both expands on and subverts previous practices; and interfaces shape the physical and gestural comportments with which we interact with material technologies and objects in the world. Procedural listening—the process of learning how to listen to and interact with music software in an experimental, speculative, and emergent manner—is the result of the shared affordances that new media platforms have introduced. When I first started researching the electronic music and interactive media cultures of Southern California in the 2010s, I had a simple question: what is sound after software? Procedural listening is one of the many possible answers to this question.

Procedural listening offers unique contributions to existing understandings of HCI. It encourages users to critically and actively reflect on their relationships with software; it contradicts technological tropes of intentionality, mastery, and control; and it inspires an understanding of digital literacy as a lifelong process of experiential learning and discovery. The tech industry of Silicon Valley is driven by an upgrade mentality in which constant innovation and technological change are valued above all else. As a result, users anxiously embrace the maximalist and consumerist aesthetics described in the first chapters of this book. By focusing the user's attention on the unique design affordances of the device, as well as the interface's capacity for experimental play, procedural listening encourages more measured approaches to music and media consumption. Further, procedural listening shifts the focus of the design process from the fixed *objects* sold on the commercial market to the *process-oriented* needs and desires of individual users and creative communities. In the end, this relational understanding of interface has the potential to provide both designers and users with a stronger sense of purpose in their interactions with technology.

By encouraging a critical opposition to the upgrade culture of Silicon Valley, procedural listening contradicts technological tropes of intentionality, mastery, and control.[27] These tropes are apparent not only in cultural practices, as in the idea of compositional intent and mastery of technique described throughout this book, but also in the driving technological force behind global capitalism—the "military-entertainment complex" (MEC). The MEC is defined by the close cooperation and sharing of resources between video game designers and the military, between movie producers and the US government, and between military propaganda and its realization in the entertainment industry more broadly.[28] Interfaces play a primary role in the shaping of the MEC, as evidenced by the use of video game controllers in the piloting of military attack drones, the use of VR technologies in soldier training, and the longer historical convergence between the internet and wartime surveillance networks. By encouraging shared agency between computers and their users, procedural listening suggests an ethical paradigm of HCI that's sensitive to the context and implications of technological use, rather than a simple tool-based, one-way relationship in which a user imposes commands on the technology.

Finally, procedural listening inspires an understanding of digital literacy as a lifelong process of experiential learning and discovery. Ray Kurzweil, an inventor and writer who has worked across the domains of music, philosophy, and computer science, once said that biology itself is a software process, claiming that humans are constantly "walking around with outdated software running in our bodies, which evolved in a very different era."[29] As computer users continually update the operating systems on their laptops or upgrade the smartphones in their pockets, they learn anew all the physical gestures and cognitive shifts required of computational systems. Following the algorithmic flow, ephemerality, and fluidity of software itself, procedural listening encourages users to embrace uncertainty and foster perennially shifting understandings of the world. Despite the apparent complexity in the software as a whole, procedural listeners understand that the moment of digital creation begins with a single line of code.

These conclusions are critical lessons for researchers and practitioners involved in sound, design, and interactive media more broadly. The hermeneutic analysis of interface design invites a new generation of scholar-practitioners to understand software as a platform in which conceptions of musicality, instrumentality, and performance are continually being negotiated. For designers, procedural listening introduces a framework with which to question long-standing practices and assumptions in the tech industry. Despite cogent arguments against the use of design metaphors from scholar-practitioners Brenda Laurel, Janet Murray, and others, the practice of designing interface elements as virtual representations of physical objects remains ever-present among professional designers. Following the experimental and abstract design examples provided throughout this book, designers may, in turn, challenge users to think outside the box in their interactions with technology. Further, user research methods among designers may be enhanced by embracing the "thick description" and context-oriented ethnographic methods of anthropologists and ethnomusicologists.[30] While some practitioners may argue that intellectual critique has no place in the design process, it's precisely the balance of theory and practice that has pushed the design and creative industries in new directions for the past hundred years.

My main goal in writing this book was to provide a critical and analytical model with which to understand my own experiences as a digital media artist working with sound, as well as the shifting landscape of media in the age of software. But planned obsolescence in the tech industry makes software especially a constantly moving target, and there will always be new areas to explore and questions to pursue. What about the role of software in global surveillance systems, for example? Wearable devices and motion capture technologies represent arguably the most dominant trends in interface design, and—as a result of their relatively recent emergence—the cultural meanings and practical uses of these technologies are still up for grabs. As such, research in global surveillance systems could provide a political and ethical angle for the discussions about tacit knowledge and embodiment present in chapters 4 and 5. In addition, conversations about race and gender dynamics within the creative industries were just heating up as I completed this book, and I look forward to more fruitful conversations between scholars and practitioners on these topics.[31] Just as the software boom of the early 2000s has supposedly democratized technological design and use, it's also sparked renewed discussions about the gender gap in the technology industry.[32] By combining ethnographic work in the technology industry with the analytical study of the software that drives the industry, future work in interface aesthetics could generate critical discussions among designers, users, and intellectuals on the topics of labor, ethics, and social justice.

It's impossible to anticipate the path that technology will follow in the coming years. In 1977, the founder of the Digital Equipment Corporation told the attendees of the World Future Society convention that "there is no reason for any individual to have a computer in his home."[33] Twenty years later, the chief technology officer of Microsoft claimed that "Apple is already dead."[34] Indeed, digital media cultures are often as fleeting and short-lived as software itself, and this makes the scholarly study and practical interrogation of these communities all the more urgent. Yet these failed predictions about our technological future only serve to support my conception of software interaction as an experimental, process-oriented endeavor in which existing conceptions of creativity, productivity, and sociocultural identity are constantly questioned. In the process of software's eternal becoming, the screens to which we dedicate so much of our physical and cognitive capacities act as both windows through which we view the world and mirrors that depict the evolving interface between our personal values, identities, and aesthetics.

Notes

Introduction

1. "Report: 90% of Waking Hours Spent Staring at Glowing Rectangles," The Onion, June 15, 2009, http://www.theonion.com/article/report-90-of-waking-hours-spent-staring-at-glowing-2747.
2. Thor Magnusson, *Sonic Writing: Technologies of Material, Symbolic, and Signal Inscriptions* (New York: Bloomsbury, 2019).
3. Shane Butler, *The Ancient Phonograph* (Cambridge, MA: MIT Press, 2015).
4. "The Human Voice Is Human on the New Orthophonic Victrola," David Sarnoff Library digital archive, March 27, 1927, http://digital.hagley.org/cdm/ref/collection/p16038coll11/id/12569.
5. "Is It Live or Is It Memorex," ukclassictelly, May 14, 2012, https://www.youtube.com/watch?v=lhfugTnXJV4.
6. Erkki Huhtamo specifically examines the rhetoric of "immediacy" and "immersion" in the nineteenth century as "topoi" that have reappeared throughout the history of interactive media and technology. *Illusions in Motion: Media Archeology of the Moving Panorama and Related Spectacles* (Cambridge, MA: MIT Press, 2013).
7. For an introduction to foley, see Patrick Winters, *Sound Design for Low and No Budget Films* (New York: Routledge, 2017).
8. Marshall McLuhan, *Understanding Media: The Extensions of Man* (New York: McGraw-Hill, 1964).
9. Jerome McGann, *Radiant Textuality: Literature after the World Wide Web* (New York: Palgrave, 2001).
10. Among designers and developers, the software "environment" is made up of the collection of resources used to support the software application, including the computer operating system, database, and development tools used to build the program. Media theorists have expanded on this integrated, networked understanding of software through the concept of "media ecology," the study of complex communication systems as environments. See Matthew Fuller, *Media Ecologies: Materialist Energies in Art and Technoculture* (Cambridge, MA: MIT Press, 2005); Christine Nystrom, "Towards a Science of Media Ecology: The Formulation of Integrated Conceptual Paradigms for the Study of Human Communication Systems," PhD diss., New York University, 1973; Neil Postman, "The Reformed English Curriculum," in *High School 1980: The Shape of the Future in American Secondary Education*, ed. Alvin C. Eurich (New York: Pitman, 1970), 160–168.
11. Mads Walther-Hansen, "New and Old User Interface Metaphors in Music Production," *Journal on the Art of Record Production* 11 (2017): n.p., https://www.arpjournal.

com/asarpwp/new-and-old-user-interface-metaphors-in-music-production/; Adam Patrick Bell, Ethan Hein, and Jarrod Ratcliffe, "Beyond Skeuomorphism: The Evolution of Music Production Software User Interface Metaphors," *Journal on the Art of Record Production* 9 (2015): n.p., https://www.arpjournal.com/asarpwp/bey ond-skeuomorphism-the-evolution-of-music-production-software-user-interface-metaphors-2/; Marianne Van Den Boomen, *Transcoding the Digital: How Metaphors Matter in New Media* (Amsterdam: Institute of Network Cultures, 2014).

12. Ben Shneiderman, "Direct Manipulation: A Step beyond Programming Languages," *Computer* 16, no. 8 (1983): 57–69.

13. The debate over "skeuomorphic" versus "flat" design is ongoing. In 2020, the conversation turned to what many referred to as "neumorphism," a blend of skeuomorphic and flat approaches. See Jack Koloskus, "Apple, Big Sur, and the Rise of Neumorphism," *Input*, June 24, 2020, https://www.inputmag.com/design/apple-macos-big-sur-the-rise-of-neumorphism; Claire L. Evans, "A Eulogy for Skeuomorphism," *Vice*, June 11, 2013, https://www.vice.com/en_us/article/nzzpyz/a-eulogy-for-skeumorphism.

14. Tom May, "The Beginner's Guide to Flat Design," *Creative Bloq*, January 14, 2021, https://www.creativebloq.com/graphic-design/what-flat-design-3132112; Amber Leigh Turner, "The History of Flat Design: How Efficiency and Minimalism Turned the Digital World Flat," *The Next Web*, March 19, 2014, http://thenextweb.com/dd/2014/03/19/history-flat-design-efficiency-minimalism-made-digital-world-flat/.

15. Google, "Material Design: Introduction," *Google*, n.d., https://material.io/design/introduction .

16. Adrian Daub, *What Tech Calls Thinking: An Inquiry into the Intellectual Bedrock of Silicon Valley* (New York: Farrar, Straus and Giroux, 2020).

17. Quoted in Michael Veal, *Dub: Soundscapes and Shattered Songs in Jamaican Reggae* (Middletown, CT: Wesleyan University Press, 2007), 42.

18. Faber, Tom. "Decolonizing Electronic Music Starts with Its Software." *Pitchfork*, February 25, 2021, https://pitchfork.com/thepitch/decolonizing-electronic-music-starts-with-its-software/.

19. Tamara Levitz, "The Musicological Elite," *Current Musicology* 102 (2018): 9–49.

20. Philip A. Ewell, "Music Theory and the White Racial Frame," *Music Theory Online* 26, no. 2 (2020), https://mtosmt.org/issues/mto.20.26.2/mto.20.26.2.ewell.html.

21. Michael Powell, "Obscure Musicology Journal Sparks Battles over Race and Free Speech," *New York Times*, February 14, 2021, https://www.nytimes.com/2021/02/14/arts/musicology-journal-race-free-speech.html.

22. Veal, *Dub*, 42.

23. Jonathan Sterne, *The Audible Past: Cultural Origins of Sound Reproduction* (Durham, NC: Duke University Press, 2003).

24. Stefan Helmreich, "An Anthropologist Underwater: Immersive Soundscapes, Submarine Cyborgs, and Transductive Ethnography," *American Ethnologist* 34 no. 4 (2007): 621–641.

25. Adrian Mackenzie (2003) defines transduction as "a way of theorizing and figuring things primarily in terms of relationality, as processes of recontextualization and in

terms of generativity." "Transduction: Invention, Innovation and Collective Life," http://www.lancs.ac.uk/staff/mackenza/papers/transduction.pdf.

26. Alexandra Supper, "The Search for the 'Killer Application': Drawing the Boundaries around the Sonification of Scientific Data," in *The Oxford Handbook of Sound Studies*, ed. Trevor Pinch and Karin Bijsterveld (New York: Oxford University Press, 2011), 249–270.

27. Karmen Franinović and Stefania Serafin, *Sonic Interaction Design* (Cambridge, MA: MIT Press, 2013).

28. Nick Montfort et al., *10 PRINT CHR$(205.5+RND(1)); : GOTO 10* (Cambridge, MA: MIT Press, 2013).

29. Lev Manovich, *Software Takes Command* (New York: Bloomsbury, 2013).

30. Friedrich Kittler, *Gramophone, Film, Typewriter*, trans. Geoffrey Winthrop-Young and Michael Wutz (Stanford, CA: Stanford University Press, 1999); Marshall McLuhan, *The Medium Is the Massage: An Inventory of Effects* (Corte Madera, CA: Gingko, 2001).

31. David Berry, *The Philosophy of Software: Code and Mediation in the Digital Age* (New York: Palgrave Macmillan, 2011); Wendy Hui Kyong Chun, *Programmed Visions: Software and Memory* (Cambridge: MIT Press, 2011); Matthew Fuller, ed., *Software Studies: A Lexicon* (Cambridge, MA: MIT Press, 2008); Casey Reas and Chandler McWilliams, *Form+Code in Design, Art, and Architecture* (New York: Princeton Architectural Press, 2010); Noah Wardrip-Fruin, *Expressive Processing: Digital Fictions, Computer Games, and Software Studies* (Cambridge, MA: MIT Press, 2009).

32. Alexander Galloway, *The Interface Effect* (Malden, MA: Polity, 2012), 23.

33. Branden Hookway, *Interface* (Cambridge, MA: MIT Press, 2014); André Nusselder, *Interface Fantasy: A Lacanian Cyborg Ontology* (Cambridge, MA: MIT Press, 2009).

34. Brenda Laurel, *Computers as Theatre* (Boston: Addison-Wesley, 1993), 116.

35. Donald Norman, *The Design of Everyday Things* (New York: Basic Books, 1988); Jacob Gube, "What Is User Experience Design? Overview, Tools and Resources," *Smashing Magazine*, October 5, 2010, http://uxdesign.smashingmagazine.com/2010/10/05/what-is-user-experience-design-overview-tools-and-resources/.

36. Janet Murray, *Inventing the Medium: Principles of Interaction Design as a Cultural Practice* (Cambridge, MA: MIT Press, 2012), 11.

37. Jason Huff, "Interface Aesthetics: An Introduction," *Rhizome*, August 3, 2012, https://rhizome.org/editorial/2012/aug/03/interface-aesthetics/.

38. Marcia Muelder Eaton, *Basic Issues in Aesthetics* (Belmont, CA: Wadsworth, 1988).

39. Immanuel Kant, *Critique of Judgment*, trans. Werner S. Pluhar (Indianapolis: Hackett, 1987).

40. Terry Eagleton, *The Ideology of the Aesthetic*, 2nd ed. (Cambridge, MA: Basil Blackwell, 1981). An overview of Eagleton's concept can be found in Carol Gigliotti, "Aesthetics of a Virtual World," *Leonardo* 28, no. 4 (1995): 290.

41. Madeleine Akrich, "The De-Scription of Technical Objects," in *Shaping Technology/Building Society: Studies in Sociotechnical Change*, ed. Wiebe E. Bijker and John Law (Cambridge, MA: MIT Press, 1992), 205–224.

42. Nina Sun Eidsheim, *Sensing Sound: Singing and Listening as Vibrational Practice* (Durham, NC: Duke University Press, 2015).

43. The work of Steven Feld and others under the rubric of "sensory ethnography" has been useful in demonstrating alternative ethnographic methods more appropriate to my stance as a scholar-practitioner. See, for example, Alex Rhys-Taylor, "The Essences of Multiculture: A Sensory Exploration of an Inner-City Street Market," *Identities* 20, no. 4 (2013): 393–406; Sarah Pink, *Doing Sensory Ethnography* (London: Sage, 2009); Tim Ingold, *The Perception of the Environment* (London: Routledge, 2000).

44. Steve Krug, *Don't Make Me Think, Revisited: A Common Sense Approach to Web and Mobile Usability* (San Francisco: New Riders, 2014); Jenifer Tidwell, *Designing Interfaces: Patterns for Effective Interaction Design* (Sebastopol, CA: O'Reilly Media, 2005); Jef Raskin, *The Humane Interface: New Directions for Designing Interactive Systems* (Boston: Addison-Wesley Professional, 2000); Russ Unger and Carolyn Chandler, *A Project Guide to UX Design: For User Experience Designers in the Field or in the Making* (Upper Saddle River, NJ: Pearson Education, 2009); Julie A. Jacko, *The Human-Computer Interaction Handbook*, 3rd ed. (Boca Raton, FL: CRC, 2012).

45. While they played crucial roles in the formation of ethnomusicology and musicology as academic disciplines, organology and sketch studies have fallen by the wayside in the past twenty years. In an attempt to understand contemporary poiesis, this book presents interface aesthetics as an analytical model that merges recent work in "critical organology" with the ontological concerns of sketch studies. These fields are increasingly significant in the digital era, as both instruments and musical materials ("sketches") move from the museum, archive, and recording studio to the hard drive. See, for example, Friedemann Sallis, ed., *Music Sketches* (Cambridge: Cambridge University Press, 2015); Eliot Bates, "The Social Life of Musical Instruments," *Ethnomusicology* 56, no. 3 (Fall 2012): 363–395; Emily Dolan, "Toward a Musicology of Interfaces," *Keyboard Perspectives* 5 (2012): 1–14.

46. Kiri Miller, *Playing Along: Digital Games, YouTube, and Virtual Performance* (New York: Oxford University Press, 2012), is exemplary in the model it creates for conducting "virtual" discourse analyses.

47. Michel Foucault, *The Archeology of Knowledge*, trans. R. Sheridan (New York: Routledge, 1972); Friedrich Kittler, *Discourse Networks 1800/1900* (Stanford, CA: Stanford University Press, 1990).

48. Tyler Cowen, "The Next Silicon Valley Could Be . . . Los Angeles?" Bloomberg, June 26, 2019, https://www.bloomberg.com/opinion/articles/2019-06-26/los-angeles-could-be-the-next-silicon-valley; Bruce Upbin, "Why Los Angeles Is Emerging as the Next Silicon Valley," *Forbes*, August 28, 2012, http://www.forbes.com/sites/ciocentral/2012/08/28/why-los-angeles-is-emerging-as-the-next-silicon-valley/.

49. *Objectified*, directed by Gary Hustwit, Plexi Productions, 2009.

50. Sherry Ortner, *Not Hollywood: Independent Film at the Twilight of the American Dream* (Durham, NC: Duke University Press, 2013), 26.

51. Peter Lunenfeld, "Mia Laboro: Maker's Envy and the Generative Humanities," Book Presence in a Digital Age Conference, University of Utrecht, 2012; Peter Lunenfeld,

"The Maker's Discourse," Critical Mass: The Legacy of Hollis Frampton, Invited Conference, University of Chicago, 2010.

52. Elisabeth Le Guin, *Boccherini's Body: An Essay in Carnal Musicology* (Berkeley: University of California Press, 2006), 5.

53. Elisabeth Le Guin, "'Cello-and-Bow Thinking': Boccherini's Cello Sonata in Eb Major, 'Fuori Catalog,'" *Echo* 1, no. 1 (1999), http://www.echo.ucla.edu/Volume1-Issue1/leguin/leguin-article.html.

54. Alan Watts, *Nature, Man, and Woman* (New York: Vintage Books, 1991).

55. Matt Pearson, *Generative Art: A Practical Guide Using Processing* (Shelter Island, NY: Manning, 2011).

56. Brian Eno, "Generative Music," *In Motion Magazine*, July 7, 1996, http://www.inmotionmagazine.com/eno1.html.

Chapter 1

1. "Las Vegas," *Noisey*, Vice, April 7, 2016.

2. Reed Jackson, "The Story of FruityLoops: How a Belgian Porno Game Company Employee Changed Modern Music," *Vice*, December 11, 2015, https://noisey.vice.com/en_us/article/rnwkvz/fruity-loops-fl-studio-program-used-to-create-trap-music-sound.

3. Dan Weiss, "The Unlikely Rise of FL Studio, the Internet's Favorite Production Software," *Vice*, October 12, 2016, https://noisey.vice.com/en_us/article/d33xzk/fl-studio-soulja-boy-porter-robinson-madeon-feature.

4. "Soundgoodizer a Blessing or a Curse?," *The FLipside*, February 22, 2009, http://www.theflipsideforum.com/index.php?topic=11536.0.

5. Peter Kirn, "How to Get into a Creative Flow with FL Studio—and What Could Make It Worth It," *CreateDigitalMusic*, June 10, 2020 https://cdm.link/2020/06/how-to-get-into-a-creative-flow-with-fl-studio/.

6. George Borzyskowski, "The Hacker Demo Scene and It's [*sic*] Cultural Artefacts," http://citeseerx.ist.psu.edu/viewdoc/download?doi=10.1.1.130.4968&rep=rep1&type=pdf; see Geert Lovink, *Dynamics of Critical Internet Culture (1994–2001)* (Amsterdam: Institute of Network Cultures, 2009), 77, for a discussion of "warezhouses" and the importance of free software for LGBTQ+ communities.

7. From Fiona Macmillan, ed., *New Directions in Copyright Law*, Vol. 1 (Northampton, MA: Edward Elgar, 2005), 230.

8. Weiss, "The Unlikely Rise."

9. David Décary-Hétu, "Police Operations 3.0: On the Impact and Policy Implications of Police Operations on the Warez Scene," *Policy and Internet* 6, no. 3 (2014): 315–340; Alf Rehn, "The Politics of Contraband: The Honor Economies of the Warez Scene," *Journal of Socio-Economics* 33 (2004): 359–374.

10. Ryan Diduck, *Mad Skills: MIDI and Music Technology in the Twentieth Century* (Marquette, MI: Repeater, 2018), 53.

11. Ian Reyes, "To Know beyond Listening: Monitoring Digital Music," *The Senses and Society* 5, no. 3 (2010): 322–338; quoted in Robert Strachan, *Sonic Technologies: Popular Music, Digital Culture, and the Creative Process* (London: Bloomsbury, 2017, 27.

12. Quoted in Stefan Goldmann, *Presets: Digital Shortcuts to Sound* (Berlin: Macro, 2019), 52.

13. Quoted in Steve Waksman, "Reading the Instrument: An Introduction," *Popular Music and Society* 26, no. 3 (2003): 252.

14. Evgeny Morozov, *To Save Everything, Click Here: The Folly of Technological Solutionism* (New York: Public Affairs, 2013).

15. Nathan Ensmenger, "Making Programming Masculine," in *Gender Codes: Why Women Are Leaving Computing*, ed. Thomas J. Misa (Hoboken, NJ: IEEE Computer Society, 2010), 115–141.

16. Adam Patrick Bell, *Dawn of the DAW: The Studio as Musical Instrument* (New York: Oxford University Press, 2018).

17. Judy Wajcman, *Feminism Confronts Technology* (Cambridge: Polity, 1991); quoted in Victoria Armstrong, "Hard Bargaining on the Hard Drive: Gender Bias in the Music Technology Classroom," *Gender and Education* 20, no. 4 (2008): 383.

18. Ann Game and Rosemary Pringle, *Gender at Work* (London: Pluto, 1984); quoted in Armstrong, "Hard Bargaining," 377.

19. Ruth Oldenzeil, *Making Technology Masculine: Men, Women, and Modern Machines in America, 1870–1945* (Amsterdam: Amsterdam University Press, 2004).

20. Paul Théberge, *Any Sound You Can Imagine: Making Music/Consuming Technology* (Hanover, NH: Welseyan University Press, 1997), 111.

21. Ibid., 123–124.

22. Steve Harvey, "Virtue and Vice and Gear Lust in Brooklyn," *Pro Sound News*, March 2018 n.p.

23. Théberge, *Any Sound*, 61.

24. Diduck, *Mad Skills*, 150.

25. Orthentix, "The Afflicting Intersections of Gender and Music Production," *Medium*, January 13, 2009, https://medium.com/orthentix/the-afflicting-intersections-of-gen der-and-music-production-a0917a41944c.

26. Georgina Born, *Rationalizing Culture: IRCAM, Boulez, and the Institutionalization of the Musical Avant-Garde* (Berkeley: University of California Press, 1995).

27. Théberge, *Any Sound*.

28. Sam, "Image Line—Jean-Marie Cannie [CTO and Co-Founder]," *Speakhertz*, May 16, 2013 http://speakhertz.com/16/interview-image-line-co-founder-jean-marie-cannie.

29. "'Fruity' seems to have all sorts of meanings associated with it that we thought best to avoid for the sake of marketing appeal." Image-Line, "Company History," http://www. image-line.com/company-history/.

30. "Listening to Music before Being a Producer/Listening to Music Being a Producer." Sam Antidote, "Top 5 Reference Tracks for Producers," *Antidote Audio*, October 19, 2017, https://www.antidoteaudio.com/blog-reference-songs.

31. Sheila Whiteley, *Women and Popular Music: Sexuality, Identity, and Subjectivity* (London: Psychology Press, 2000); M. Bayton, "Women and the Electric

Guitar," in *Sexing the Groove: Popular Music and Gender*, ed. Sheila Whiteley (New York: Routledge, 1997), 37–49; S. A. O'Neill and M. J. Boulton, "Boys' and Girls' Preferences for Musical Instruments: A Function of Gender?," *Psychology of Music* 24, no. 2 (1996): 171–183; Judith K. Delzell and David A. Leppla, "Gender Association of Musical Instruments and Preferences of Fourth-Grade Students for Selected Instruments," *Journal of Research in Music Education* 40, no. 2 (1992): 93–103; R. D. Crowther and K. Durkin, "Sex- and Age-Related Differences in the Musical Behavior, Interests and Attitudes towards Music of 232 Secondary School Students," *Educational Studies* 8, no. 2 (1982): 131–139.

32. Armstrong, "Hard Bargaining," 384.

33. Lucy Green, *Music, Gender, Education* (Cambridge: Cambridge University Press, 1997), 84.

34. Simon Reynolds, "Maximal Nation: Electronic Music's Evolution toward the Thrilling Excess of Digital Maximalism," *Pitchfork*, December 6, 2011, http://pitchfork.com/features/articles/8721-maximal-nation/.

35. Ibid.

36. Matthew Ingram, "Switched-On," *Loops* 1, 2009, 136.

37. Ibid.

38. Quoted in Jackson, "The Story of FruityLoops."

39. Ibid.

40. Ulrik Ekman, ed., *Throughout: Art and Culture Emerging with Ubiquitous Computing* (Cambridge, MA: MIT Press, 2012).

41. Fuller, *Media Ecologies*.

42. William Powers, *Hamlet's BlackBerry: Building a Good Life in the Digital Age* (New York: Harper Perennial, 2011).

43. Alex Hunley, "Hamlet's BlackBerry—William Powers," *Full Stop*, August 24, 2011, http://www.full-stop.net/2011/08/24/reviews/alexhunley/hamlets-blackberry-william-powers/.

44. Nicholas Carr, *The Shallows: What the Internet Is Doing to Our Brains* (New York: W. W. Norton, 2011).

45. Frenchy Lunning, ed., *Mechademia 5: Fanthropologies* (Minneapolis: University of Minnesota Press, 2010), 140.

46. Richard H. R. Harper, *Texture: Human Expression in the Age of Communications Overload* (Cambridge, MA: MIT Press, 2010).

47. Tiziana Terranova, *Network Culture: Politics for the Information Age* (London: Pluto, 2004), 1.

48. Ibid.

49. Waksman, "Reading the Instrument."

50. Vlad Georgescu, "Feature Creep Is a Problem. Learn How to Avoid It," *Plutora*, May 12, 2019, https://www.plutora.com/blog/feature-creep-problem.

51. Tiziana Terranova, "Free Labor: Producing Culture for the Digital Economy," *Social Text* 18, no. 2 (2000): 33–58.

52. Lucie Vágnerová, "'Nimble Fingers' in Electronic Music: Rethinking Sound through Neo-colonial Labour," *Organised Sound* 22, no. 2 (2018): 250–258.

53. Théberge, *Any Sound.*

54. Rachel Laine, "Syn & Synthesizers," Flickr.com, https://www.flickr.com/photos/bifr ostgirl/albums/72157649002021637.

55. James McNally, "Favela Chic: Diplo, *Funk Carioca*, and the Ethics and Aesthetics of the Global Remix," *Popular Music and Society* 40, no. 4 (2016): 434–452.

56. "Diplo about Native Instruments KOMPLETE," Native Instruments, October 19, 2010, https://www.youtube.com/watch?v=9j7Aor_4NTE.

57. Boima Tucker, "Global Genre Accumulation," *Africa Is a Country*, November 22, 2011, https://africasacountry.com/2011/11/global-genre-accumulation/.

58. "How Diplo Is Making Pop Music Weird," *Time*, August 20, 2015, https://www.yout ube.com/watch?v=lxRDCsYAX50.

59. Tara Rodgers, "Synthesizing Sound: Metaphor in Audio-Technical Discourse and Synthesis History," PhD diss., McGill University, 1. Also see Greene 2005, 5–6; Théberge 2004; Reiffenstein 2006.

60. Native Instruments, "Discovery Series: Collection," https://www.native-instruments. com/en/products/komplete/world/discovery-series-collection/.

61. Robin James, *Resilience & Melancholy: Pop Music, Feminism, Neoliberalism* (Alresford: Zero Books, 2015), 98.

62. Ibid., 103–104.

63. David Bessell, "Blurring Boundaries: Trends and Implications in Audio Production Software Developments," in *The Oxford Handbook of Interactive Audio*, ed. Karen Collins, Bill Kapralos, and Holly Tessler (New York: Oxford University Press, 2014), 408.

64. Miller Puckette, "Max at Seventeen," *Computer Music Journal* 26, no. 4 (2002): 40.

65. Quoted in Goldmann, *Presets*, 71, 76.

66. James, *Resilience & Melancholy*, 5.

67. This refers to the following claim: "Society uses aesthetics in order to control us through our buying patterns, and to coerce us into buying higher priced commodities. . . . Therefore, it's a subversive act to rediscover and value that which is cheap and readily available . . . what society has thrown out." V. Vale and Andrea Juno, *Incredibly Strange Music*, Vol. 1 (San Francisco: RE/Search, 1993), 4; cited in Strachan, *Sonic Technologies*, 147.

68. Ableton, "Machinedrum: Sacred Frequencies," September 25, 2012, https://www.able ton.com/en/blog/machinedrum/.

69. Joe Muggs, "Flying Lotus: Cosmic Drama," *Resident Advisor*, March 24, 2010, http:// www.residentadvisor.net/feature.aspx?1175.

70. Ableton, "What Is This Book?" https://makingmusic.ableton.com/about.

71. Dennis DeSantis, *Making Music: 74 Creative Strategies for Electronic Music Producers* (Berlin: Ableton, 2015), 58.

72. Ibid., 79.

73. Ibid., 59.

74. Timothy Taylor, *The Sounds of Capitalism: Advertising, Music, and the Conquest of Culture* (Chicago: University of Chicago Press, 2012), 4.

75. Peter McAllister, *Manthropology: The Science of Why the Modern Male Is Not the Man He Used to Be* (New York: St. Martin's, 2010).

76. Pankaj Mishra, "The Crisis in Modern Masculinity," *Guardian*, March 17, 2018, https://www.theguardian.com/books/2018/mar/17/the-crisis-in-modern-masculinity.

Chapter 2

1. Taking inflation into account, $995 in 1990 had the same buying power as about $2,000 in 2021. In contrast, the average price of a full-fledged DAW in 2021 fell to between $100 and $400.

2. Adam Patrick Bell, "DAW Democracy? The Dearth of Diversity in 'Playing the Studio,'" *Journal of Music, Technology & Education* 8, no. 2 (2015): 129–146; Timothy Taylor, "The Commodification of Music at the Dawn of the Era of 'Mechanical Music,'" *Ethnomusicology* 51, no. 2 (2007): 281–305.

3. Armstrong, "Hard Bargaining"; G. Folkestad, D. J. Hargreaves, and B. Lindstrom, "A Typology of Composition Styles," *British Journal of Music Education* 15, no. 1 (1998): 83–97; Richard Hodges, "The New Technology," in *Music Education: Trends and Issues*, ed. Charles Plummeridge (London: Institute of Education, University of London, 1996), XXX–XXX.

4. S. Kiesler, L. Sproull, and J. S. Eccles, "Pool Halls, Chips and War Games: Women in the Culture of Computing," *Psychology of Women Quarterly* 9 (1985): 451–462; Hank Bromley and Michael W. Apple, eds., *Education/Technology/Power: Educational Computing as Social Practice* (Albany: State University of New York Press, 1998); S. Clegg, "Theorising the Machine: Gender, Education and Computing," *Gender and Education* 13, no. 3 (2001): 307–324; L. Stepulevage, "Gender/Technology Relations: Complicating the Gender Binary," *Gender and Education* 13, no. 3 (2001): 325–338; A. Colley and C. Comber, "Age and Gender Differences in Computer Use and Attitudes among Secondary School Students: What Has Changed?" *Educational Research* 45, no. 2 (2003): 155–165; Sherry Turkle, *The Second Self: Computers and the Human Spirit* (New York: Simon & Shuster, 1984).

5. Martin Heidegger, "The Question Concerning Technology," in *Basic Writings*, ed. David Farrell Krell (New York: Harper & Row, 1977), 287.

6. Jay David Bolter and Richard Grusin, *Remediation: Understanding New Media* (Cambridge, MA: MIT Press, 2000), 65.

7. Quoted in Bell, *Dawn of the DAW*, 84.

8. Bell, *Dawn of the DAW*, 84.

9. Ibid.

10. Erich M. von Hornbostel and Curt Sachs, "Classification of Musical Instruments: Translated from the Original German by Anthony Baines and Klaus P. Wachsmann," *Galpin Society Journal* 14 (1961): 3–29.

11. Margaret Kartomi, "The Classification of Musical Instruments: Changing Trends in Research from the Late Nineteenth Century, with Special Reference to the 1990s," *Ethnomusicology* 45, no. 2 (2001): 305.

12. Quoted in Bell, *Dawn of the DAW*, 35.

13. I'm aware of the implementation of non-"musical" technologies and devices as musical instruments (washboards, radio static, vinyl records, and more), but I'm here addressing instruments that were designed explicitly *as* instruments.

14. For example, Théberge characterizes the theremin as an instrument that bore no resemblance to any existing musical technology, thus requiring musicians not only to adapt to unfamiliar sounds but also to learn an entirely new set of performance techniques. *Any Sound*, 44.

15. Van Den Boomen, *Transcoding the Digital*, XXX–XXX.

16. Mark Butler surveys how "work-concepts" have been employed across a range of musical discourses as a means of asserting the compositional authority, originality, and boundedness of musical objects. *Playing with Something That Runs: Technology, Improvisation, and Composition in DJ and Laptop Performance* (New York: Oxford University Press, 2014). See also Georgina Born, "On Musical Mediation: Ontology, Technology, and Creativity," *Twentieth-Century Music* 2, no. 1 (2005): 7–36; Albin J. Zak, *The Poetics of Rock: Cutting Tracks, Making Records* (Berkeley: University of California Press, 2001); Lydia Goehr, *The Imaginary Museum of Musical Works: An Essay in the Philosophy of Music* (Oxford: Oxford University Press, 1992); Theodore Gracyk, *Rhythm and Noise: An Aesthetics of Rock* (Durham, NC: Duke University Press, 1996).

17. Ann Light, "Performing Interaction Design with Judith Butler," in *Critical Theory and Interaction Design*, ed. Jeffrey Bardzell, Shaowen Bardzell, and Mark Blythe (Cambridge, MA: MIT Press, 2018), 438.

18. Lars-Erik Janlert and Erik Stolterman, *The Things That Keep Us Busy: The Elements of Interaction* (Cambridge: MIT Press, 2017), 119.

19. Ibid., 134.

20. I'm using the term in quite a different way here, but a detailed look at how "professionalism" became such an important facet of neoliberal capitalism can be found in Magali Sarfatti Larson, *The Rise of Professionalism: Monopolies of Competence and Sheltered Markets* (Livingston, NJ: Transaction, 2012).

21. Ruha Benjamin, *Race after Technology: Abolitionist Tools for the New Jim Code* (Cambridge: Polity, 2019), 179.

22. "Is It Time for Avid to Refresh the Pro Tools Graphical User Interface?—Poll," *Pro Tools Expert*, May 13, 2019, https://www.pro-tools-expert.com/home-page/2019/4/25/is-it-time-for-avid-to-overhaul-the-entire-pro-tools-user-interface-for-the-future.

23. Melissa Mandelbaum, "Applying Architecture to Product Design: Parti," *Medium*, March 21, 2016, https://medium.com/applying-architecture-to-product-design/applying-architecture-to-product-design-parti-68b0b7345db9#.p0d87v6lc.

24. Matthew Frederick, *101 Things I Learned in Architecture School* (Cambridge: MIT Press, 2007).

25. Ibid.

26. Quoted in Bell, "DAW Democracy?," 139.

27. Jesse!, "FL Studio Is . . . Compared to Pro Tools?," Image-Line forum, February 28, 2012, https://forum.image-line.com/viewtopic.php?f=100&t=89594; mdb, "You Don't Need Pro Tools Dude!," Ableton forum, June 18, 2006, https://forum.ableton.com/viewtopic.php?t=40683.

28. Soumavo De Sarkar, "What Is special or Different about Pro Tools vs Any Other DAW? What Actually Makes It Industry Standard?," *Quora*, June 15, 2020, https://www.quora.com/What-is-special-or-different-about-Pro-Tools-vs-any-other-DAW-What-actually-makes-it-industry-standard.

29. Quoted in Armstrong, "Hard Bargaining," 377. See also Marcia J. Citron, *Gender and the Musical Canon* (Cambridge: Cambridge University Press, 1993), 52. Finally, James makes this point in the context of EDM in *Resilience & Melancholy*.

30. Tami Gadir, "Resistance or Reiteration? Rethinking Gender in DJ Cultures," *Contemporary Music Review* 35, no. 1 (2016): 120.

31. Kyle Vorbach, "Ben Shapiro Produces EDM EP Using Logic and Reason," *The Hard Times*, July 12, 2019, https://thehardtimes.net/culture/ben-shapiro-produces-edm-ep-using-logic-and-reason/.

32. Bell, "DAW Democracy?," 136.

33. Wajcman, *Feminism Confronts Technology*.

34. Georgina Born and Kyle Devine, "Introduction: Gender, Creativity and Education in Digital Musics and Sound Art," *Contemporary Music Review* 35, no. 1 (2016): 6–7. See also Green, *Music, Gender, Education*; Lucy Green, "Music Education, Cultural Capital, and Social Group Identity," in *The Cultural Study of Music: A Critical Introduction*, ed. Martin Clayton, Trevor Herbert, and Richard Middleton (New York: Routledge), 206–216; Victoria Armstrong, *Technology and the Gendering of Music Education* (Farnham, UK: Ashgate, 2011); Robert Legg, "'One Equal Music': An Exploration of Gender Perceptions and the Fair Assessment by Beginning Music Teachers of Musical Compositions," *Music Education Research* 12, no. 2 (2010): 141–149.

35. Armstrong, *Technology*, 119.

36. Ibid., 8.

37. Shaowen Bardzell, "Through the 'Cracks and Fissures' in the Smart Home to Ubiquitous Utopia," in *Critical Theory and Interaction Design*, ed. Jeffrey Bardzell, Shaowen Bardzell, and Mark Blythe (Cambridge: MIT Press, 2018), 774.

38. While the music products industry is always focusing on the next big thing, it's important to consider—following Wendy Hui Kyong Chun—the importance of creative media practices that have become habitual (rather than simply "innovative" or "new"). For Chun, this is how media becomes embedded in our daily lives, and this is why it can be difficult to take a critical stance on the objects that have come to define who we are. Wendy Hui Kyong Chun, *Updating to Remain the Same: Habitual New Media* (Cambridge, MA: MIT Press, 2016).

39. Quoted in Allie Barbera, "Professor, Songwriter, Producer, Musician, Music Technology Consultant, and Now Soon-to-Be Mama: Erin Barra-Jean," *JSJEvents*,

October 23, 2017, https://www.jsjeventsboston.com/blog/2017/9/25/getting-to-know-erin-barra-founder-of-beatz-by-girls.

40. Ableton's website continually archives blog posts and artist interviews. Huckaby's quote was found at Ableton, "Artist Quotes," Ableton, n.d., https://www.ableton.com/en/pages/artists/artist_quotes/.

41. "Angélica Negrón: Toys Noise," Ableton, January 13, 2012, https://www.ableton.com/en/blog/angelica-negron/.

42. "Clint Sand: Electronic Polymath," Ableton, January 20, 2011, https://www.ableton.com/en/pages/artists/clint_sand/.

43. "Sowall: From Jazz Chops to Finger Drumming," Ableton, October 1, 2019, https://www.ableton.com/en/blog/sowall-jazz-chops-finger-drumming/.

44. "Laura Escudé: The Art of Designing a Live Show," Ableton, September 24, 2019, https://www.ableton.com/en/blog/laura-escude-art-designing-live-show/.

45. "Sakura Tsuruta on Push," Ableton, October 21, 2019, https://www.ableton.com/en/blog/sakura-tsuruta-push/.

46. "Suzi Analogue: Staying in the Zone," Ableton, March 8, 2018, https://www.ableton.com/en/blog/suzi-analogue-staying-zone/.

47. Steven Johnson, *Interface Culture* (New York: HarperCollins, 1997), 84.

48. Mark Katz, *Capturing Sound: How Technology Has Changed Music* (Berkeley: University of California Press, 2004).

49. David W. Bernstein, *The San Francisco Tape Music Center: 1960s Counterculture and the Avant-Garde* (Berkeley: University of California Press, 2008).

50. Earl Morrogh, *Information Architecture: An Emerging 21st Century Profession* (Saddle River, NJ: Prentice Hall, 2003); Louis Rosenfeld and Peter Morville, *Information Architecture for the World Wide Web* (Sebastopol, CA: O'Reilly, 1998); Richard Saul Wurman, *Information Architects* (New York: Graphis, 1997).

51. Wurman, *Information Architects*, 17.

52. Andrea Resmini and Luca Rosati, *Pervasive Information Architecture: Designing Cross-Channel User Experiences* (Burlington, MA: Morgan Kauffman, 2011).

53. Christina Wodtke, *Information Architecture: Blueprints for the Web* (San Francisco: New Riders, 2009); Peter Brown, *Information Architecture with XML* (Hoboken, NJ: John Wiley, 2003).

54. Wei Ding and Xia Lin, *Information Architecture: The Design and Integration of Information Spaces* (San Rafael, CA: Morgan & Claypool, 2009).

55. Magnusson, *Sonic Writing*, 52.

56. Van Den Boomen, *Transcoding the Digital*, 12.

57. Jason Stanyek and Benjamin Piekut, "Deadness: Technologies of the Intermundane," *Drama Review* 54, no. 1 (Spring 2010): 20.

58. Paul Théberge, "The Network Studio: Historical and Technological Paths to a New Ideal in Music Making," *Social Studies of Science* 34, no. 5 (2004): 776.

59. Manovich, *Software Takes Command*, 156.

60. Frances Dyson, *Sounding New Media: Immersion and Embodiment in the Arts and Culture* (Berkeley: University of California Press, 2009), 3.

61. Alan Kay, "Computer Software," *Scientific American* 251, no. 3 (1984): 52–59.

62. Manovich, *Software Takes Command*, 142.

63. Peter Lunenfeld, *The Secret War between Downloading and Uploading: Tales of the Computer as Culture Machine* (Cambridge, MA: MIT Press, 2011), 28.

64. Many media theorists and computer scientists employ architectural analogies in describing the "environment" of software applications. Johnson interrogates the cultural and political import of these metaphors, claiming that "each [software] design decision echoes and amplifies a set of values, an assumption about the larger society that frames it. All works of architecture imply a worldview, which means that all architecture is in some deeper sense political" (*Interface Culture*, 44).

65. Following Walter Benjamin's argument about photography in "The Work of Art in the Age of Mechanical Reproduction," Arild Bergh and Tia DeNora examine how recording and distribution technologies "have potentially democratised the field of aesthetic music experiences" and erased "the line between listener/fan and record producer/patron." For a comprehensive overview of the term "democratization" as it's been used in strains of social and cultural theory, see Patryk Galuszka, "Netlabels and Democratization of the Recording Industry," *First Monday* 17, no. 7 (July 2012): n.p., http://firstmonday.org/article/view/3770/3278.

66. Valerie Bauman, "78% of Women in the Music Industry Say They're Treated Differently at Work Because of Their Gender, Study Says," *Daily Mail*, March 12, 2019, https://www.dailymail.co.uk/news/article-6801023/78-Women-music-industry-say-theyre-treated-differently-work-gender.html; Anastasia Tsioulcas, "A Year after the #MeToo Grammys, Women Are Still Missing in Music," NPR, February 8, 2019, https://www.npr.org/2019/02/08/692671099/a-year-after-the-metoo-grammys-women-are-still-missing-in-music.

67. Homepage, Beats By Girlz, https://beatsbygirlz.org.

68. Freida Abtan, "Where Is She? Finding the Women in Electronic Music Culture," *Contemporary Music Review* 35, no. 1 (2016): 58.

69. Sarah Benet-Weiser, *Empowered: Popular Feminism and Popular Misogyny* (Durham, NC: Duke University Press, 2018; Robin James, "Poptimism and Popular Feminism," *Sounding Out!*, September 17, 2018, https://soundstudiesblog.com/2018/09/17/poptimism-and-popular-feminism/.

70. Born and Devine, "Introduction"; Benjamin, *Race after Technology*, 61.

71. Leslie Gaston-Bird, *Women in Audio* (London and New York: Routledge, 2020), xiv.

72. Quoted in Benjamin, *Race after Technology*, 19.

73. Paula Wolfe, "A Studio of One's Own: Music Production, Technology, and Gender," *Journal on the Art of Record Production* 7 (2012): n.p., https://www.arpjournal.com/asarpwp/a-studio-of-one's-own-music-production-technology-and-gender/.

74. Shaowen Bardzell, "Feminist HCI: Taking Stock and Outlining an Agenda for Design," in *Proceedings of the SIGCHI Conference on Human Factors in Computing Systems* (New York: Association for Computing Machinery, 2010), 1307.

75. Benjamin, *Race after Technology*, 176.

76. Ibid., 176, 179.

77. Jace Clayton, *Uproot: Travels in 21st-Century Music and Digital Culture* (New York: Farrar, Straus and Giroux, 2016), 177–178.

78. Mark Blythe, "After Critical Design," in *Critical Theory and Interaction Design*, ed. Jeffrey Bardzell, Shaowen Bardzell, and Mark Blythe (Cambridge, MA: MIT Press, 2018), 507.
79. Ibid., 186–187.
80. Ibid., 192.
81. Ibid., 188.
82. Milena Radzikowska et al., Jennifer Roberts-Smith, Xinyue Zhou, and Stan Ruecker, "A Speculative Feminist Approach to Project Management," *Strategic Design Research Journal* 12, no. 1 (2019): 94–113.
83. Miller Puckette, "The Deadly Embrace between Music Software and Its Users," in *Electroacoustic Music beyond Performance: Proceedings of the Electroacoustic Music Studies Network Conference* Berlin: EMS Network, 2014), XXX–XXX.
84. Ibid.

Chapter 3

1. Puckette named *Max/MSP* in honor of computer music pioneer Max Mathews. What he refers to as the *Max* paradigm includes three computer programs: *Max/MSP*, *Jmax*, and *Pure Data* (*Pd*). For more information, see Puckette, "Max at Seventeen."
2. Cycling '74, the company that sells *Max*, specifically described the software as "a full kit of creative tools for sound, graphics, music and interactivity in a visual environment" in 2016. In 2020, it changed the description to "software for experimentation and invention." See https://cycling74.com.
3. From the organization's website: "Anyone, anywhere can organize an Hour of Code event. One-hour tutorials in over 45 languages. No experience needed." Code.org, "Hour of Code," http://hourofcode.com/us. See also Microsoft, "Get Your Start with an Hour of Code," http://www.microsoft.com/about/corporatecitizenship/en-us/youthspark/youthsparkhub/hourofcode/.
4. Henry Jenkins, *Confronting the Challenges of Participatory Culture: Media Education for the 21st Century* (Cambridge, MA: MIT Press, 2009).
5. Doug Belshaw, *The Essential Elements of Digital Literacies* (self-published, 2014), https://dougbelshaw.com/essential-elements-book.pdf.
6. M. Sharples et al., "Maker Culture: Learning by Making," in *Innovating Pedagogy 2013: Open University Innovation Report 2* (Milton Keynes, UK: Open University, 2013), 33–35.
7. Ilias Bergstrom and R. Beau Lotto, "Code Bending: A New Creative Coding Practice," *Leonardo* 48, no. 1 (2015): 25–31; Eduardo Ledesma, "The Poetics and Politics of Computer Code in Latin America: Codework, Code Art, and Live Coding," *Revista de Estudios Hispánicos* 49, no. 1 (2015): 91–120; Alex McLean, Julian Rohrhuber, and Nick Collins, eds., Special Issue on Live Coding, *Computer Music Journal* 38, no. 1 (2014); Nick Collins, "Live Coding of Consequence," *Leonardo* 44, no. 3 (2011): 207–211; Thor Magnusson, "Algorithms as Scores: Coding Live Music," *Leonardo Music Journal* 21 (2011): 19–23.

8. "Livecoding.tv: Watch People Code Products Live," Education Ecosystem Blog, February 9, 2015, https://blog.education-ecosystem.com/livecoding-tv-watch-peo ple-code-products-live/.

9. Chapter 4 discusses in more detail historical and contemporary debates surrounding live performance with computers.

10. Christopher M. Kelty, *Two Bits: The Cultural Significance of Free Software* (Durham, NC: Duke University Press, 2008).

11. Born, *Rationalizing Culture*, 184.

12. Ibid., 233.

13. While code appears to hold close affinity to natural language, Born notes that programming languages are "far from transparent to decode, even for the highly skilled authors themselves. However, IRCAM programmers seemed to delight in this intransigent opacity since, despite the many difficulties that it caused, it made programs appear artful and unstandardized expressions of collective imaginative labor." Ibid., 230.

14. Ibid., 284.

15. Miller Puckette, "The Patcher," in *Proceedings of the International Computer Music Association* (San Francisco: ICMA, 1988), 420–429. http://msp.ucsd.edu/Publicati ons/icmc88.pdf.

16. Puckette describes the Macintosh computer that was brought to IRCAM in 1987 by David Wessel, "without whose efforts I and the rest of IRCAM might have entirely missed out on the personal computer revolution." "Max at Seventeen," 34.

17. "Ableton and Cycling '74 Form a New Partnership," Ableton, June 7, 2017, https:// www.ableton.com/en/blog/ableton-cycling-74-new-partnership/.

18. David Zicarelli, "An Interview with William Kleinsasser," Cycling '74, September 13, 2005, https://cycling74.com/2005/09/13/an-interview-with-william-kleinsasser/.

19. Andrew Pask, "Mini Interview: Jonny Greenwood," Cycling '74, January 2, 2014, https://cycling74.com/2014/01/02/mini-interview-jonny-greenwood/.

20. Editing parameters at the "micro" level of an audio sample are possible as a result of music software that allows for high-resolution digital sampling in the first place. Similar techniques are outlined in Paul Harkins, "Microsampling: From Akufen's Microhouse to Todd Edwards and the Sound of UK Garage," in *Musical Rhythm in the Age of Digital Reproduction*, ed. Anne Danielsen (Burlington, VT: Ashgate, 2012), 177–194.

21. Andrew Spitz, "Skube: A Last.fm & Spotify Radio," September 12, 2012, https://vimeo. com/50435491.

22. Jeanette Wing, "Computational Thinking," *Communications of the ACM* 49, no. 3 (2006): 33–35.

23. Gena R. Greher and Jesse M. Heines, *Computational Thinking in Sound: Teaching the Art and Science of Music and Technology* (New York: Oxford University Press, 2014), 4.

24. Manovich, *Software Takes Command*; Berry, *The Philosophy of Software*; Chun, *Programmed Visions*; Reas and McWilliams, *Form+Code*; Wardrip-Fruin, *Expressive Processing*; Fuller, *Software Studies*.

25. Michael Mateas, "Procedural Literacy: Educating the New Media Practitioner," in *Beyond Fun: Serious Games and Media*, ed. Drew Davidson (Pittsburgh,: ETC, 2008),

67–83, http://citeseerx.ist.psu.edu/viewdoc/download?doi=10.1.1.90.2106&rep= rep1&type=pdf.

26. Ian Bogost, "Persuasive Games: The Proceduralist Style," *Gamasutra*, January 21, 2009, http://www.gamasutra.com/view/feature/132302/persuasive_games_the_.php.

27. Ian Bogost, *Unit Operations: An Approach to Videogame Criticism* (Cambridge, MA: MIT Press, 2006), 3.

28. Chun, *Programmed Visions*, 225.

29. Bogost, "Persuasive Games."

30. Wardrip-Fruin, *Expressive Processing*, 7.

31. Butler defines this position of being both performer and observer as "listener orientation," defined within the specific context of EDM: "This term captures a widespread set of attitudes within electronic dance music. A DJ or laptop set characterized by listener orientation is simultaneously performance-based and interpretive; it encompasses both the production and consumption of sound." Butler, *Playing with Something*, 106.

32. Steve Reich, "Music as a Gradual Process," in *Writings on Music, 1965–2000* (New York: Oxford University Press, 2004), 34.

33. Eno, "Generative Music," n.p.

34. Ben Nevile, "An Interview with Kim Cascone," Cycling '74, September 13, 2005, https://cycling74.com/2005/09/13/an-interview-with-kim-cascone/. Emphasis added.

35. The writings and legacy of John Cage remain the strongest advocates of an experimental musical practice in opposition to compositional intention and control. He notes that "those involved with the composition of experimental music find ways and means to remove themselves from the activities of the sounds they make." John Cage, *Silence: Lectures and Writings* (Hanover, NH: Wesleyan University Press, 1961), 10.

36. Sam Tarakajian, "An Interview with Ducky," Cycling '74, November 18, 2013, https://cycling74.com/2013/11/18/an-interview-with-ducky/.

37. Norman, *The Design of Everyday Things*.

38. Ibid., 8.

39. Ibid., 12.

40. Ibid., 9. Also see James J. Gibson, "Theory of Affordances," in *The Ecological Approach to Visual Perception* (Hillsdale, NJ: Lawrence Erlbaum, 1986).

41. Apple, "Designing for iOS," in *iOS Human Interface Guidelines* (Cupertino, CA: Apple, 2014), https://developer.apple.com/library/iOS/documentation/userexperience/conceptual/mobilehig/.

42. Manovich, *Software Takes Command*, 100.

43. See the work of researchers affiliated with the International Conference on New Interfaces for Musical Expression (NIME). For example, Thor Magnusson, "Affordances and Constraints in Screen-Based Musical Instruments," in *Proceedings of the 4th Nordic Conference on Human-Computer Interaction: Changing Roles* (New York: ACM, 2006), 441–444; Atau Tanaka, Alessandro Altavilla, and Neal Spowage, "Gestural Musical Affordances," in *Proceedings of the 9th Sound and Music Computing Conference* (Copenhagen: SMC, 2012), 318–325; Atau Tanaka, "Mapping Out Instruments, Affordances, and Mobiles," in *Proceedings of the 2010 Conference*

on New Interfaces for Musical Expression (Sydney: NIME, 2010), 15–18. In addition to NIME, various articles from the archives of *Leonardo Music Journal, Journal of New Music Research*, and *Computer Music Journal* contain topics related to interface design.

44. Joseph Butch Rovan, "Living on the Edge: Alternate Controllers and the Obstinate Interface," in *Mapping Landscapes for Performance as Research: Scholarly Acts and Creative Cartographies*, ed. Shannon Rose Riley and Lynette Hunter (New York: Palgrave Macmillan, 2009), 253.

45. Chris Nash and Alan F. Blackwell, "Flow of Creative Interaction with Digital Music Notations," in *The Oxford Handbook of Interactive Audio*, ed. Karen Collins, Bill Kapralos, and Holly Tessler (New York: Oxford University Press, 2013), 387–404.

46. Dolan, "Toward a Musicology of Interfaces," 11.

47. Pask, "Mini Interview."

48. Nancy Mauro-Flude, "The Intimacy of the Command Line," *Scan* 10, no. 2 (2013): n.p., http://scan.net.au/scn/journal/vol10number2/Nancy-Mauro-Flude.html.

49. In outlining the elements of procedural listening throughout this chapter, I simultaneously build on and push against concepts and terminology that have become staples of "design thinking" and the neoliberal Silicon Valley ("Californian") ideology (iteration, proceduralism, rapid prototyping, process, etc.). This is not to occlude other musical histories and cultural antecedents such as the long histories of experimental rock and jazz music that have employed process-based creative workflows, improvisational compositional practices, and other forms of iterative "prototyping." Most important, I use the terms here to explain how software-centric musical cultures simultaneously reify and push against the potentially problematic ideals that have defined the tech industries for so long.

50. David Zicarelli, "How I Learned to Love a Program That Does Nothing," *Computer Music Journal* 26, no. 4 (2002): 44.

51. Wardrip-Fruin, *Expressive Processing*, 7.

52. Nevile, "An Interview."

53. Quoted in Frank Dietrich, "Visual Intelligence: The First Decade of Computer Art (1965–1975)," *Leonardo* 19, no. 2 (1986): 166.

54. Emily Grace Adiseshiah, "Iterative Prototyping to Improve the Design Process," *Justinmind*, January 14, 2016, http://www.justinmind.com/blog/iterative-prototyping-to-improve-the-design-process/.

55. Brad Larson, "Rapid Prototyping in Swift Playgrounds," *objc.io* 16 (September 2014): n.p., https://www.objc.io/issues/16-swift/rapid-prototyping-in-swift-playgrounds/.

56. Elisabeth Skjelten, *Complexity and Other Beasts* (Oslo, Norway: The Oslo School of Architecture and Design, 2014). See also Birger Sevaldson's website, "Systems Oriented Design," Oslo School of Architecture and Design, accessed August 28, 2020, http://www.systemsorienteddesign.net.

57. Manovich, *Software Takes Command*, 46.

58. Nevile, "An Interview."

59. Wardrip-Fruin, *Expressive Processing*, 13–14.

60. Again, I'm foregrounding concepts from computer science to push against their established meanings. While the "network" studio ideal builds on previous conceptions of how pro audio gear might be used, networks are also defined by boundaries established by the limitations of their constituents. These boundaries are reflected and embodied in the relatively homogeneous demographic makeup of so many musical cultures born from the *Max* paradigm and other code-based audio tools (live coding, electroacoustic music, some branches of "controllerism," etc.).

61. Ezra Mechaber, "President Obama Is the First President to Write a Line of Code," White House Blog, December 10, 2014, http://www.whitehouse.gov/blog/2014/12/10/president-obama-first-president-write-line-code.

62. David Golumbia, *The Cultural Logic of Computation* (Cambridge, MA: Harvard University Press, 2009).

63. Tara McPherson, "Why Are the Digital Humanities So White? Or Thinking the Histories of Race and Computation," in *Debates in the Digital Humanities*, ed. Matthew K. Gold (Minneapolis: University of Minnesota Press, 2012), 139–160.

64. Eric S. Raymond, *The Art of Unix Programming* (Boston: Addison-Wesley, 2003), https://nakamotoinstitute.org/static/docs/taoup.pdf.

65. McPherson, 148.

66. Clive Thompson, *Coders: The Making of a New Tribe and the Remaking of the World* (London: Penguin, 2019).

67. Eric Roberts, "A History of Capacity Challenges in Computer Science," Stanford Computer Science, March 7, 2016, https://cs.stanford.edu/people/eroberts/CSCapacity.pdf.

68. Kelty, *Two Bits*, 29.

69. Ibid., 38.

70. Stewart Brand, "Spacewar," *Rolling Stone*, December 7, 1972, http://www.wheels.org/spacewar/stone/rolling_stone.html.

71. Nathan Ensmenger, *The Computer Boys Take Over: Computers, Programmers, and the Politics of Technical Expertise* (Cambridge, MA: MIT Press, 2010), 57.

72. Ibid., 43.

73. Erika Lorraine Milam and Robert A. Nye, "An Introduction to *Scientific Masculinities*," *Osiris* 30, no. 1 (2015): 5.

74. Ibid., 14.

75. See the discussion forum thread by Titus Bellwald, "OT Are There Maxing Women Out There," Cycling '74, June 3, 2006, https://cycling74.com/forums/ot-are-there-maxing-women-out-there. "Are there any women on this list? It seems like a men's club to me, am I wrong? Is *Max* like math or other sciences that are dominated by males?"

Chapter 4

1. These concerns echo the 1985 court case initiated by the Parents Music Resource Center (PMRC) against commercial music with violent and sexual content. U.S. Senate, "Record Labeling: Hearing before the Committee on Commerce, Science,

and Transportation," Ninety-Ninth Congress, First Session on Contents of Music and the Lyrics of Records, September 19, 1985, http://www.joesapt.net/superlink/shrg99-529/. In the context of EDM, see Rong-Gong Lin II, "ER Doctors: Drug-Fueled Raves Too Dangerous and Should Be Banned," *Los Angeles Times*, August 10, 2015, http://www.latimes.com/local/lanow/la-me-ln-why-some-er-doctors-want-to-end-raves-in-los-angeles-county-20150810-story.html; Ben Sisario and James C. McKinley Jr., "Drug Deaths Threaten Rising Business of Electronic Music Fests," *New York Times*, September 9, 2013, http://www.nytimes.com/2013/09/10/arts/music/drugs-at-music-festivals-are-threat-to-investors-as-well-as-fans.html?_r=0. In the context of video games, see Heather Kelly, "Do Video Games Cause Violence?" CNN, August 17, 2015, http://money.cnn.com/2015/08/17/technology/video-game-violence/; Brett Molina, "Obama Seeks Research into Violent Video Games," *USA Today*, January 16, 2013, http://www.usatoday.com/story/tech/gaming/2013/01/16/obama-gun-violence-video-games/1839879/.

2. "Benga Makes a Grime Beat on PlayStation Music 2000—BBC 2003," GetDarker, December 5, 2014, https://www.youtube.com/watch?v=AZHnaSIZP0Y; "Skream Talks Dubstep Warz, Hatcha and FWD," Red Bull Music Academy, September 24, 2011, https://youtu.be/wqchdQQV_QY.

3. "Diggin' in the Carts, Series Trailer," Red Bull Music Academy, August 28, 2014, https://www.youtube.com/watch?v=fwN_o_fi7xE.

4. Mark Katz, *Groove Music: The Art and Culture of the Hip-Hop DJ* (New York: Oxford University Press, 2012), 64.

5. Tricia Rose, *Black Noise: Rap Music and Black Culture in Contemporary America* (Hanover, NH: Wesleyan University Press, 1994), 89.

6. Ibid., 78.

7. Katz, *Groove Music*, 154–155.

8. Joseph G. Schloss, *Making Beats: The Art of Sample-Based Hip-Hop* (Middletown, CT: Wesleyan University Press, 2004), 80.

9. Katz, *Groove Music*, 218.

10. Sarah Thornton, *Club Cultures: Music, Media, and Subcultural Capital* (Hanover, NH: Wesleyan University Press, 1996), 11.

11. Brian Barrett, "End of an Era: Panasonic Kills Off Technics Turntables," *Gizmodo*, October 28, 2010, http://gizmodo.com/5675818/end-of-an-era-panasonic-kills-off-technics-turntables. The SL-1200 was rereleased in 2016, marketed to an audience of mostly audiophiles rather than DJs. Nilay Patel, "The Technics SL-1200 Turntable Returns in Two New Audiophile Models," *The Verge*, January 5, 2016, http://www.theverge.com/2016/1/5/10718234/technics-sl1200-sl1200g-sl1200gae-turntable-new-models-announced-release-date-ces-2016.

12. Quoted in Ed Montano, "'How Do You Know He's Not Playing Pac-Man While He's Supposed to Be DJing?': Technology, Formats and the Digital Future of DJ Culture," *Popular Music* 29 no. 3 (2010): 410. This issue was the target of satire in Andy Samberg's *Saturday Night Live* skit "When Will the Bass Drop?," NBC, May 18, 2014, https://youtu.be/DoUV7Q1C1SU.

13. It's important to note at this point that not all "live" settings for EDM are the same. While some audiences value the perceived "musical" skill or technical virtuosity being demonstrated by the DJ, others prioritize the social experience of dancing and other

nightlife rituals that occur in the performance spaces of EDM. For a different reading of the relationship between dance music technologies, remix practices, and the concept of "liveness," see Micah Salkind, *Do You Remember House? Chicago's Queer of Color Undergrounds* (New York: Oxford University Press, 2019).

14. Deadmau5, "We All Hit Play," *United We Fail*, 2013, http://deadmau5.tumblr.com/post/25690507284/we-all-hit-play. Also see DJ A-Trak's perspective: "Don't Push My Buttons," *Huffington Post,* July 23, 2012, accessed August 28, 2020, http://www.huffing tonpost.com/atrak/dont-push-my-buttons_b_1694719.html.

15. Again, I'm using the term "audience members" to refer to individuals at EDM events who tend to dedicate their time assessing the perceived musicality of the stage performance rather than dancing. I understand that the line between "audience members" and "crowd of dancers" is always blurry within the multimedia spectacle of EDM.

16. See, for example, Rebekah Farrugia and Thomas Swiss, "Tracking the DJs: Vinyl Records, Work, and the Debate over New Technologies," *Journal of Popular Music Studies* 17, no. 1 (2005): 33. Raymond Williams describes the historical dialectic that occurs in the negotiation of new technologies: "At first glance there are simply dire predictions based on easily aroused prejudices against new technologies. Yet there are also phases of settlement in which formerly innovating technologies have been absorbed and only the currently new forms are a threat." Raymond Williams, *Television: Technology and Cultural Form*, 3rd ed. (New York: Routledge, 2003), 133.

17. Butler, *Playing with Something*, 96.

18. "Moldover: Performance and Controllerism," Ableton, November 28, 2013, https://www.ableton.com/en/blog/moldover-performance-and-controllerism/.

19. Diduck, *Mad Skills*, 148.

20. Quoted in Mark Dery, "Black to the Future: Interviews with Samuel R. Delany, Greg Tate, and Tricia Rose," in *Flame Wars: The Discourse of Cyberculture*, ed. Mark Dery (Durham, NC: Duke University Press, 1994), 192.

21. Alexander G. Weheliye, "Rhythms of Relation: Black Popular Music and Mobile Technologies," *Current Musicology* 99–100 (2017): 107.

22. Quoted in Veal, *Dub*, 42.

23. For more information on the historical development of the MIDI protocol, see Jessica Feldman, "The MIDI Effect," paper presented at Bone Flute to Auto-Tune: A Conference on Music & Technology in History, Theory and Practice, Berkeley, CA, April 24–26, 2014.

24. Native Instruments, "Maschine Studio," February 1, 2016, http://www.native-inst ruments.com/en/products/maschine/production-systems/maschine-studio/.

25. Quoted in Ean Golden, "Music Maneuvers: Discover the Digital Turntablism Concept, 'Controllerism,' Compliments of Moldover," *Remix*, October 2007, http://www.moldover.com/press/Moldover_Remix_Oct-2007_w.jpg.

26. Quoted in Brian Kane, *Sound Unseen: Acousmatic Sound in Theory and Practice* (New York: Oxford University Press, 2014), 16.

27. Quoted in Daniela Cascella, "Sound Objects: Pierre Schaeffer's 'In Search of a Concrete Music,'" *Los Angeles Review of Books*, April 3, 2013, http://lareviewofbo

oks.org/review/sound-objects-pierre-schaeffers-in-search-of-a-concrete-music. Emphasis added.

28. Peggy Phelan, *Unmarked: The Politics of Performance* (New York: Routledge, 1993), 146.

29. Carolyn Abbate, "Music—Drastic or Gnostic," *Critical Inquiry* 30, no. 3 (2004): 506.

30. Charles Keil, "The Theory of Participatory Discrepancies: A Progress Report," *Ethnomusicology* 39, no. 1 (1995): 3.

31. Ibid., 13.

32. Philip Auslander, *Liveness: Performance in a Mediatized Culture*, 2nd ed. (New York: Routledge, 2008), 10.

33. For example, see Boris Kachka, Rebecca Milzoff, and Dan Reilly, "10 Tricks That Musicians and Actors Use in Live Performances," *Vulture*, April 3, 2016, http://www. vulture.com/2016/03/live-performance-tricks.html; Chenda Ngak, "Tupac Coachella Hologram: Behind the Technology," CBS News, November 9, 2012, http://www.cbsn ews.com/news/tupac-coachella-hologram-behind-the-technology/.

34. Simon Emmerson, *Living Electronic Music* (Burlington, VT: Ashgate, 2007), 27.

35. Ibid., 53.

36. Primus Luta, "Toward a Practical Language for Live Electronic Performance," *Sounding Out!*, April 29, 2013, http://soundstudiesblog.com/2013/04/29/toward-a-practical-language-for-live-electronic-performance/.

37. Primus Luta, "Musical Objects, Variability and Live Electronic Performance," *Sounding Out!*, August 12, 2013, http://soundstudiesblog.com/2013/08/12/musical-objects-variability/.

38. Robin James, "Aesthetics and Live Electronic Music Performance," *Cyborgology*, December 13, 2013, http://thesocietypages.org/cyborgology/2013/12/13/aesthetics-and-live-electronic-music-performance/.

39. Quoted in Bates, "The Social Life," 372.

40. Ibid., 364.

41. Butler, *Playing with Something*, 4.

42. Joost Raessens, *Homo Ludens 2.0: The Ludic Turn in Media Theory* (Utrecht: Utrecht University, 2012); Joost Raessens, "Computer Games as Participatory Media Culture," in *Handbook of Computer Game Studies*, ed. Joost Raessens and Jeffrey Goldstein (Cambridge, MA: MIT Press, 2006), 373–388.

43. Louis Pattison, "Video Game Loops & Aliens: Flying Lotus Interviewed," *The Quietus*, May 18, 2010, http://thequietus.com/articles/04269-flying-lotus-interview-cosm ogramma. Flying Lotus's studio setup, which includes "a mess of keyboards, DVDs, video games, computers, and a drum kit," attests to this phenomenon of media convergence. Jeff Weiss, "Flying Lotus' Nocturnal Visions," *LA Weekly*, October 4, 2012, http://www.laweekly.com/music/flying-lotus-nocturnal-visions-2611706.

44. Drew Millard, "Interview: Rustie Used to Produce Like How Gamers Game," *Pitchfork*, May 31, 2012, http://archive.is/dzPoJ.

45. Roger Moseley, "Playing Games with Music (and Vice Versa): Ludomusicological Perspectives on *Guitar Hero* and *Rock Band*," in *Taking It to the Bridge: Music as*

Performance, ed. Nicholas Cook and Richard Pettengill (Ann Arbor: University of Michigan Press, 2013), 286.

46. Jane McGonigal, *Reality Is Broken: Why Games Make Us Better and How They Can Change the World* (New York: Penguin, 2011); Tom Bissell, *Extra Lives: Why Video Games Matter* (New York: Pantheon Books, 2010); Katie Salen and Eric Zimmerman, *Rules of Play: Game Design Fundamentals* (Cambridge, MA: MIT Press, 2003).

47. Bernard Suits, *The Grasshopper: Games, Life, and Utopia* (Peterborough, ON: Broadview, 2005), 55.

48. The marketing of music technology as allowing an unfettered creative experience is discussed in depth in Théberge, *Any Sound.*

49. Brian Upton, *The Aesthetic of Play* (Cambridge, MA: MIT Press, 2015), 33–34.

50. Harmony Bench, "Gestural Choreographies: Embodied Disciplines and Digital Media," in *The Oxford Handbook of Mobile Music Studies*, Vol. 2, ed. Sumanth Gopinath and Jason Stanyek (New York: Oxford University Press, 2014), 238.

51. Ibid., 243.

52. Ibid., 245.

53. See also Ingrid Richardson, "Faces, Interfaces, Screens: Relational Ontologies of Framing, Attention and Distraction," *Transformations* 18 (2010): http://www.transformationsjournal.org/wp-content/uploads/2017/01/Richardson_Trans18.pdf; Ingrid Richardson, "Ludic Mobilities: The Corporealities of Mobile Gaming," *Mobilities* 5 (2010): 431–447; Ingrid Richardson and Rowan Wilken, "Haptic Vision, Footwork, Place-Making: A Peripatetic Phenomenology of the Mobile Phone Pedestrian," *Second Nature* 1 (2009): 22–41; Ian MacColl and Ingrid Richardson, "A Cultural Somatics of Mobile Media and Urban Screens: Wiffiti and the IWALL Prototype," *Journal of Urban Technology* 15, no. 3 (2008): 99–116; Ingrid Richardson, "Pocket Technospaces: The Bodily Incorporation of Mobile Media," *Continuum* 21 (2007): 205–215; Ingrid Richardson, "Mobile Technosoma: Some Phenomenological Reflections on Itinerant Media Devices," *Fibreculture Journal* 6 (2005): https://six.fibreculturejournal.org/fcj-032-mobile-technosoma-some-phenomenological-reflections-on-itinerant-media-devices/; Rowan Wilken, "From *Stabilitas Loci* to *Mobilitas Loci:* Networked Mobility and the Transformation of Place," *Fibreculture Journal* 6 (2005): https://six.fibreculturejournal.org/fcj-036-from-stabilitas-loci-to-mobilitas-loci-networked-mobility-and-the-transformation-of-place/.

54. David Myers, "The Video Game Aesthetic: Play as Form," in *Video Game Theory Reader 2*, ed. Bernard Perron and Mark J. P. Wolf (New York: Routledge, 2009), 50.

55. For more information on the technical design and use of gaming controllers, see Pippin Barr, James Noble, and Robert Biddle, "Video Game Values: Human-Computer Interaction and Games," *Interacting with Computers* 19 (2007): 180–195; Ben Heatherly and Logan Howard, "Video Game Controllers," Clemson University, http://andrewd.ces.clemson.edu/courses/cpsc414/spring14/papers/group2.pdf.

56. Myers, "The Video Game Aesthetic," 46.

57. Le Guin, "Cello-and-Bow Thinking."

58. Michael Polanyi, *The Tacit Dimension* (Chicago: University of Chicago Press, 1966).

59. Butler describes the "adaptability" inherent in the Monome's generality of function, as "the buttons have no set correspondence with particular musical parameters or

function; indeed, the interface need not control music at all." Butler, *Playing with Something*, 88.

60. Bart Simon, "Wii Are Out of Control: Bodies, Game Screens and the Production of Gestural Excess," March 5, 2009, https://papers.ssrn.com/sol3/papers.cfm?abstract_id=1354043.

61. Butler, *Playing with Something*, 101.

62. Ibid., 3.

63. "Daedelus, Live at Ground Kontrol: Far from Home #9." Intothewoods.tv, July 9, 2012, https://www.youtube.com/watch?v=QY3mHBj_dhs.

64. "Ander's Incredible Custom Controller for Live," Ableton, July 13, 2012, https://www.ableton.com/en/blog/anders-incredible-custom-controller-live/.

65. "Decap and Brady Watt on Push & Working with Ski Beatz," Ableton, April 16, 2014, https://www.ableton.com/en/blog/decap-and-brady-watt-push-working-ski-beatz/.

66. Tobias Van Veen and Bernardo Alexander Attias, "Off the Record: Turntablism and Controllerism in the 21st Century (Part 1)," *Dancecult* 3, no. 1 (2011), https://dj.dancecult.net/index.php/dancecult/article/view/319.

67. Kim Cascone, "The Aesthetics of Failure: 'Post-Digital' Tendencies in Contemporary Computer Music," *Computer Music Journal* 24, no. 4 (2000): 13.

68. Krug, *Don't Make Me Think*.

69. Friedrich Kittler, "There Is No Software," in *The Truth of the Technological World: Essays on the Genealogy of Presence* (Stanford, CA: Stanford University Press, 2014), 219–229.

70. Alexander G. Weheliye, *Phonographies: Grooves in Sonic Afro-Modernity* (Durham, NC: Duke University Press, 2005).

71. Katherine McKittrick and Alexander G. Weheliye, "808s & Heartbreak," *Propter Nos* 2, no. 1 (2017): 13–42.

72. Peter Kirn, "Live Music Tools at the Bleeding Edge: Laura Escudé, from Herbie Hancock Collabs to Solo (Live Berlin, Sat)," *CDM*, July 27, 2012, https://cdm.link/2012/07/live-music-tools-at-the-bleeding-edge-laura-escude-from-herbie-hancock-collabs-to-solo-live-berlin-sat/.

73. Johan Huizinga, *Homo Ludens: A Study of the Play-Element in Culture* (Boston: Beacon, 1955), 10.

74. Bonnie Ruberg, *Video Games Have Always Been Queer* (New York: New York University Press, 2019), 309.

75. Jonathan Bollen, "Queer Kinesthesia: Performativity on the Dance Floor," in *Dancing Desires: Choreographing Sexualities On and Off the Stage*, ed. Jane C. Desmond (Madison: University of Wisconsin Press, 2001), 285–314.

76. Kiri Miller, *Playable Bodies: Dance Games and Intimate Media* (New York: Oxford University Press, 2017), 3.

Chapter 5

1. As Timothy Taylor notes, the term "democratization" was used by player piano advertisers in the early twentieth century and has been widely discussed in music and

technology literature. Following Taylor, I use the term throughout this chapter both as an accepted vernacular concept and as an ideological tool used by software companies for the purpose of advertising. Taylor, "The Commodification of Music." See also Lisa Gitelman, "Media, Materiality, and the Measure of the Digital: Or, the Case of Sheet Music and the Problem of Piano Rolls," in *Memory Bytes: History, Technology, and Digital Culture*, ed. Lauren Rabinovitz and Abraham Geil (Durham, NC: Duke University Press, 2004), 199–217; Théberge, *Any Sound*; Craig H. Roell, *The Piano in America, 1890–1940* (Chapel Hill: University of North Carolina Press, 1989).

2. Timothy D. Taylor, Mark Katz, and Tony Grajeda, eds., *Music, Sound, and Technology in America: A Documentary History of Early Phonograph, Cinema, and Radio* (Durham, NC: Duke University Press, 2012).

3. Taylor, "The Commodification of Music," 289.

4. Ibid., 291.

5. Mark Katz, "The Amateur in the Age of Mechanical Music," in *The Oxford Handbook of Sound Studies*, ed. Trevor Pinch and Karin Bijsterveld (New York: Oxford University Press, 2012), 459–479; Katz, *Capturing Sound*; Miller, *Playing Along*.

6. Ensmenger, *The Computer Boys*; J. C. R. Licklider, "Man-Computer Symbiosis," *IRE Transactions on Human Factors in Electronics* 1 (1960): 4–11; Norbert Wiener, "Men, Machines, and the World About," in *Medicine and Science*, ed. I. Galderston (New York: International Universities Press, 1954), 13–28; Alan Turing, "Computing Machinery and Human Intelligence," *Mind* 59, no. 236 (1950): 433–460.

7. Paul E. Ceruzzi, *Computing: A Concise History* (Cambridge, MA: MIT Press, 2012), 23–48; James Gleick, *The Information: A History, A Theory, A Flood* (New York: Pantheon Books, 2011); Rich Arzoomanian, "A Complete History of Mainframe Computing," *Tom's Hardware*, June 26, 2009, https://www.tomshardware.com/picturestory/508-mainframe-computer-history.html.

8. Roy A. Allan, *A History of the Personal Computer: The People and the Technology* (London, ON: Allan, 2001), https://archive.org/details/A_History_of_the_Personal_Computer; Steven Levy, *Hackers: Heroes of the Computer Revolution* (Garden City, NY: Doubleday, 1984).

9. Steve Wozniak, *iWoz* (New York: W. W. Norton, 2006), 150.

10. "1984 Apple's Macintosh Commercial (HD)," Mac History, February 1, 2012, https://www.youtube.com/watch?v=VtvjbmoDx-I.

11. "Steve Jobs Introduces iMovie & iMac DV: Apple Special Event (1999)," Dailymotion, n.d., https://www.dailymotion.com/video/x3m5log.

12. Jean Burgess, "Vernacular Creativity and New Media," PhD diss., University of Queensland, 2007.

13. Jean Burgess, "The iPhone Moment, the Apple Brand, and the Creative Consumer: From 'Hackability and Usability' to Cultural Generativity," in *Studying Mobile Media: Cultural Technologies, Mobile Communication, and the iPhone*, ed. Larissa Hjorth, Jean Burgess, and Ingrid Richardson (New York: Routledge, 2012), 37–38.

14. "Steve Jobs Introduces Rainbow iMacs & Power Mac G3: Macworld SF (1999)," Dailymotion, n.d., https://www.dailymotion.com/video/x3mssqj.

15. "The computer offers its users a formal system, but it is also active and interactive. It is easily anthropomorphized. Its experts do not think that it is 'alive.' But it is a medium onto which lifelike properties can be easily projected. It supports the fantasy 'that there is somebody home.' It is, of course, only a machine, but because of its psychological properties it supports an experience with it as an 'intimate machine.'" Sherry Turkle, "Computational Reticence: Why Women Fear the Intimate Machine," in *Technology and Women's Voices*, ed. Cheris Kramarae (New York: Pergamon, 1986), 43.

16. Ibid. See also "The Incredible, Amazing, Awesome Apple Keynote," justanotherguy84, September 15, 2009, accessed August 28, 2020, https://www.youtube.com/watch?v=Nx7v815bYUw.

17. "Making the iPhone: The Design Tao of Apple's Jony Ive," Bloomberg, September 18, 2013, https://www.youtube.com/watch?v=e9OJMlVQBiY.

18. *Objectified*, Hustwit.

19. Apple, "iOS Design Themes," in *Human Interface Guidelines*, https://developer.apple.com/design/human-interface-guidelines/ios/overview/themes/.

20. Upon the release of the app in 2015, users were scored based on the tightness of their performance with a prerecorded track, and they could upload their high scores for a chance to perform Reich's composition onstage with the London Sinfonietta.

21. Robert J. K. Jacob et al., "Reality-Based Interaction: A Framework for Post-WIMP Interfaces," in *CHI 2008 Proceedings: Post-WIMP* (Florence: ACM, 2008), 201.

22. Ingrid Richardson, "Touching the Screen: A Phenomenology of Mobile Gaming and the iPhone," in *Studying Mobile Media: Cultural Technologies, Mobile Communication, and the iPhone*, ed. Larissa Hjorth, Jean Burgess, and Ingrid Richardson (New York: Routledge, 2012), 144.

23. Edward Tufte, "iPhone Resolution," November 25, 2008, https://www.youtube.com/watch?v=YslQ2625TR4.

24. The idea of music and media production as an experience rather than a tool-based endeavor is common in literature on "post-digital" aesthetics. Kim Cascone foreshadowed many of these trends in "The Aesthetics of Failure." See also Christopher Haworth, "Sound Synthesis Procedures as Texts: An Ontological Politics in Electroacoustic and Computer Music," *Computer Music Journal* 39, no. 1 (2015): 41–58.

25. Søren Bro Pold and Christian Ulrik Andersen, "Controlled Consumption Culture: When Digital Culture Becomes Software Business," in *The Imaginary App*, ed. Paul D. Miller and Svitlana Matviyenko (Cambridge, MA: MIT Press, 2014), 17.

26. The resulting "fealty of stillness and sameness" in mobile media interactions is described in Aden Evens, "Touch in the Abstract," *SubStance* 70, no. 3 (2011): 70.

27. Robert Rosenberger, "The Spatial Experience of Telephone Use," *Environment, Space, Place* 2, no. 2 (2010): 61–75.

28. "Figure for iPhone and iPad," Reason Studios, April 4, 2012, https://www.youtube.com/watch?v=gLLjRH6GJec.

29. Sherry Turkle, *Alone Together: Why We Expect More from Technology and Less from Each Other* (New York: Basic Books, 2011).

30. Jesper Juul, *A Casual Revolution: Reinventing Video Games and Their Players* (Cambridge, MA: MIT Press, 2010), 37.

31. Apple, "Use Multi-Touch Gestures on Your Mac," n.d., https://support.apple.com/en-us/HT204895.

32. Apple, "Switch apps on Your iPhone, iPad, or iPod touch," December 1, 2020, https://support.apple.com/en-us/HT202070.

33. Mark Weiser, "The Computer for the 21st Century," *Scientific American* 265, no. 3 (1991): 94.

34. Ibid.

35. Robert Hamilton, Jeffrey Smith, and Ge Wang, "Social Composition: Musical Data Systems for Expressive Mobile Music," *Leonardo Music Journal* 21 (2011): 64.

36. Apple, "GarageBand," n.d., http://www.apple.com/ios/garageband/.

37. Reason Studios, "Figure," https://www.reasonstudios.com/mobile-apps.

38. Quoted in Jason Farman, *Mobile Interface Theory: Embodied Space and Locative Media* (New York: Routledge, 2012), 11.

39. The icon-grid interface and mobile media more broadly reflect "platformativity: a gradual shift away from the Internet as a single hub for web-based cultures, towards a multiplicity of platforms including 'cloud'-based software, social media, tablets, smartphones, and 'app'-based interfaces." Joss Hands, "Introduction: Politics, Power, and 'Platformativity,'" *Culture Machine* 14 (2013): 1–9.

40. Jonathan Zittrain, *The Future of the Internet—And How to Stop It* (New Haven, CT: Yale University Press, 2008).

41. Morozov, *To Save Everything*. See also Jean Baudrillard on the idea of the "gadget," in William Merrin, "The Rise of the Gadget and Hyperludic Me-dia," *Cultural Politics* 10, no. 1 (2014): 8.

42. Svitlana Matviyenko, "Introduction," in *The Imaginary App*, ed. Paul D. Miller and Svitlana Matviyenko (Cambridge, MA: MIT Press, 2014), xxvii.

43. Ibid., xvii. See also "iPhone 3G Commercial 'There's an App for That' 2009," CommercialsKid, February 4, 2009, https://www.youtube.com/watch?v=szrsfeyLzyg.

44. Apple, "Creativity Apps: iOS," n.d., http://www.apple.com/si/creativity-apps/ios/.

45. Ibid.

46. Quoted in Chris Butler, *Henri Lefebvre: Spatial Politics, Everyday Life and the Right to the City* (New York: Routledge, 2012), 29.

47. Pold and Andersen, "Controlled Consumption Culture," 23–24.

48. Taylor, *The Sounds of Capitalism*, 4.

49. Google, "The Basics of Micro-Moments," https://www.thinkwithgoogle.com/consumer-insights/consumer-journey/micro-moments-understand-new-consumer-behavior/.

50. Rebecca Borison, "The 15 Highest-Grossing iPhone and iPad Games," *Business Insider*, May 20, 2014, http://www.businessinsider.com/highest-grossing-iphone-and-ipad-games-2014-5?op=1.

51. Joost Raessens, "Homo Ludens 2.0: The Ludic Turn in Media Theory," address at Utrecht University, November 19, 2010, http://dspace.library.uu.nl/handle/1874/255181.

52. "About Us," n.d., https://allihoopa.com/about?&_suid=145919376448608643549 13122952.

53. "What the iPhone lacks in technological 'hackability,' therefore, it makes up for in social and cultural 'generativity,' thanks to its usability and the proliferation of apps that extend its functionality, and more importantly thanks to the creative, social and communicative activities of its millions of users who have integrated the iPhone, as a platform, into their everyday lives." Burgess, "The iPhone Moment," 40.

54. Merrin, "The Rise of the Gadget," 16.

55. Lawrence Lessig, *Remix: Making Art and Commerce Thrive in the Hybrid Economy* (New York: Penguin, 2008); Aram Sinnreich, *Mashed Up: Music, Technology, and the Rise of Configurable Culture* (Amherst: University of Massachusetts Press, 2010).

56. Taylor, *The Sounds of Capitalism*, 239.

57. Ge Wang, "Ocarina: Designing the iPhone's Magic Flute," *Computer Music Journal* 38, no. 2 (2014): 10.

58. Ibid., 20.

59. Carlo Vercellone, "Wages and Rent: The New Articulation of Wages, Rent and Profit in Cognitive Capitalism," Generation Online, n.d., https://www.generation-online.org/c/fc_rent2.htm; Yann Moulier-Boutang, *Cognitive Capitalism* (Boston: Polity, 2011).

60. Nick Dyer-Witheford, "App Worker," in *The Imaginary App*, ed. Paul D. Miller and Svitlana Matviyenko (Cambridge, MA: MIT Press, 2014), 135.

61. Vágnerová, " 'Nimble Fingers,' " 251.

Chapter 6

1. Barr, Noble, and Biddle, "Video Game Values."

2. Juul, *A Casual Revolution*.

3. "Game mechanics" refers to the "rule-based systems that facilitate and encourage a user to explore and learn the properties of their possibility space through the use of feedback mechanisms." In other words, they are the affordances that structure gameplay for the player. Daniel Cook, "What Are Game Mechanics?" *Lostgarden*, October 23, 2006, http://www.lostgarden.com/2006/10/what-are-game-mechanics.html.

4. Useful surveys of the relationship between indie game aesthetics and the oppositional attitudes of indie culture can be found in Nadav Lipkin, "Examining Indie's Independence: The Meaning of 'Indie' Games, the Politics of Production, and Mainstream Co-optation," *Loading* 7, no. 11 (2012): 8–24; Paolo Ruffino, "Independent Games: Narratives of Emancipation in Video Game Culture," *Loading* 7, no. 11 (2012): 106–121; Orlando Guevara-Villalobos, "Cultures of Independent Game Production: Examining the Relationship between Community and Labour," in *Think Design Play: Proceedings of the 2011 Digital Games Research Association Conference* (Utrecht: DIGRA, 2011); Chase Bowen Martin and Mark Deuze, "The Independent Production of Culture: A Digital Games Case Study," *Games and Culture* 4, no. 3 (2009): 276–295.

5. "Sound design" can be a vague term among digital media artists. In this chapter, I use it to describe how gameplay mechanics combine with the diegetic and non-diegetic musical elements of the overall game in constructing a holistic sonic experience informed by the dual layers of design and practice. This approach reflects conventional wisdom in sound studies scholarship, which views sound design as a social force made up of shared knowledges, practices, and cultural imaginations. See Jonathan Sterne, "Sonic Imaginations," in *The Sound Studies Reader*, ed. Jonathan Sterne (New York: Routledge, 2012), 1–18.

6. Lev Manovich, *The Language of New Media* (Cambridge, MA: MIT Press, 2001).

7. N. Katherine Hayles, *How We Became Posthuman: Virtual Bodies in Cybernetics, Literature, and Informatics* (Chicago: University of Chicago Press, 1999).

8. Nicolas Bourriaud, *Relational Aesthetics* (Dijon: Les Presses du Réel, 1998).

9. Nick Montfort and Ian Bogost, *Racing the Beam: The Atari Video Computer System* (Cambridge, MA: MIT Press, 2009).

10. For example, *Call of Duty: Black Ops III* sold $550 million in its opening weekend (2015).

11. Mike D'Errico, "Going Hard: Bassweight, Sonic Warfare, and the 'Brostep' Aesthetic," *Sounding Out!*, January 23, 2014, https://soundstudiesblog.com/2014/01/23/going-hard-bassweight-sonic-warfare-the-brostep-aesthetic/.

12. Megan Farokhmanesh, "Minecraft Console Sales Surpass PC, 'Almost 54M' Copies Sold in Total," *Polygon*, June 25, 2014, http://www.polygon.com/2014/6/25/5843358/minecraft-console-sales-54m-copies-sold.

13. Craig Stern, "What Makes a Game Indie: A Universal Definition," *Sinister Design*, August 22, 2012, http://sinisterdesign.net/what-makes-a-game-indie-a-universal-definition/.

14. Gabriela T. Richard and Kishonna L. Gray, "Gendered Play, Racialized Reality: Black Cyberfeminism, Inclusive Communities of Practice and the Intersections of Learning, Socialization and Resilience in Online Gaming," *Frontiers* 39, no. 1 (2018): 114.

15. "E3 2011 Machinima Coverage—Sound Shapes on the PS VITA Game Demo & Interview," *Machinima*, June 9, 2011, https://www.youtube.com/watch?v=_rRU0bNeJ20.

16. "Fez," *Polytron*, http://fezgame.com.

17. Jack Halberstam, *The Queer Art of Failure* (Durham, NC: Duke University Press, 2011).

18. Mihai Nadin, "Emergent Aesthetics: Aesthetic Issues in Computer Arts," *Leonardo* 2 (1989): 43–48. Scholars have introduced many terms to describe audio in gameplay that exists as more than simply a sonic backdrop to the overall media experience, including "procedural," "dynamic," "adaptive," "interactive," and "generative," among others; see Karen Collins, Bill Kapralos, and Holly Tessler, eds., *The Oxford Handbook of Interactive Audio* (New York: Oxford University Press, 2013), for a broad overview. Throughout this chapter, I use the adjective "emergent" to encompass the features and aesthetic effects shared by these terms, including nonlinearity of narrative, goal-less exploration, system components that respond and react to user input in real time, and abstraction of audiovisual content.

19. Glen Carlson, "Emergent Aesthetics," http://www.glencarlson.com/emergent/Car lson_EmergentAesthetics.pdf.

20. Phosfiend Systems, "Fract," http://fractgame.com.

21. Twisted Tree Games, "Proteus," http://twistedtreegames.com/proteus/.

22. Karen Collins, *Playing with Sound: A Theory of Interacting with Sound and Music in Video Games* (Cambridge, MA: MIT Press, 2013), 39.

23. William Cheng, *Sound Play: Video Games and the Musical Imagination* (Oxford: Oxford University Press, 2014); Miller, *Playing Along.*

24. Andy Farnell, "Procedural Audio Theory and Practice," in *The Oxford Handbook of Interactive Audio,* ed. Karen Collins, Bill Kapralos, and Holly Tessler (New York: Oxford University Press, 2014), XXX–XXX.

25. Karen Collins, "An Introduction to Procedural Music in Video Games," *Contemporary Music Review* 28, no. 1 (2009): 7.

26. Niels Böttcher, Héctor P. Martinez, and Stefania Serafin, "Procedural Audio in Computer Games Using Motion Controllers: An Evaluation on the Effect and Perception," *International Journal of Computer Games Technology* XX (2013): XXX–XXX.

27. Bogost, "Persuasive Games."

28. Dietrich, "Visual Intelligence," 162.

29. Mitchell Whitelaw, "System Stories and Model Worlds: A Critical Approach to Generative Art," in *Readme 100: Temporary Software Art Factory,* ed. Olga Goriounova (Norderstedt: BoD, 2005), 3.

30. Peter Cariani, "Emergence and Artificial Life," in *Artificial Life II: SFI Studies in the Sciences of Complexity,* ed. Christopher Langton et al. (New York: Addison-Wesley, 1991), 775–798; Christopher Langton, "Artificial Life," in *Artificial Life: SFI Studies in the Sciences of Complexity,* ed. Christopher Langton (New York: Addison-Wesley, 1989), 1–47; Martin Gardner, "Mathematical Games: The Fantastic Combinations of John Conway's New Solitaire Game 'Life,'" *Scientific American* 223, no. 4 (1970): 120–123; Turing, "Computing Machinery."

31. Kelsey Atherton, "Microsoft Is Training AI in Minecraft," *Popular Science,* March 14, 2016, http://www.popsci.com/microsoft-is-training-ai-in-minecraft.

32. Rene Wooller et al., "A Framework for Comparison of Processes in Algorithmic Music Systems," in *Proceedings from Generative Arts Practice* (Sydney: Creativity and Cognition Studios Press, 2005), 109–124.

33. Tae Hong Park, "An Interview with Max Mathews," *Computer Music Journal* 33, no. 3 (2009): 9–22.

34. Puckette, "Max at Seventeen."

35. Although he doesn't explicitly use the term "emergence," Williams outlines technological cause and effect as a two-way dialectic rather than a linear progression. Williams, *Television, Technology and Cultural Form.*

36. Bruno Latour, *Reassembling the Social: An Introduction to Actor-Network-Theory* (New York: Oxford University Press, 2005); Bruno Latour, "Technology Is Society Made Durable," in *A Sociology of Monsters? Essays on Power, Technology and Domination,* ed. John Law (London: Routledge, 1991), 103–131; Michel Callon

and Bruno Latour, "Unscrewing the Big Leviathan: How Actors Macrostructure Reality and How Sociologists Help Them to Do So," in *Advances in Social Theory and Methodology: Toward an Integration of Micro- and Macro-Sociologies*, ed. K. D. Knorr-Cetina and A. V. Cicourel (Boston: Routledge and Kegan Paul, 1981), 277–303.

37. Allucquère Rosanne Stone, *The War of Desire and Technology at the Close of the Mechanical Age* (Cambridge, MA: MIT Press, 1995).

38. Márcio Ribeiro, Paulo Borba, and Claus Brabrand, *Emergent Interfaces for Feature Modularization* (Cham, Switzerland: Springer International, 2014).

39. Penny Sweetser, *Emergence in Games* (Boston: Course Technology, 2008), 178.

40. This idea expands on the work of media scholars in the subdisciplines of interface studies and post-digital aesthetics. David M. Berry and Michael Dieter, eds., *Postdigital Aesthetics: Art, Computation and Design* (New York: Palgrave Macmillan, 2015); Hookway, *Interface*; Christian Ulrik Andersen and Søren Bro Pold, eds., *Interface Criticism: Aesthetics beyond Buttons* (Aarhus: Aarhus University Press, 2011).

41. Norbert Herber, "The Composition-Instrument: Emergence, Improvisation, and Interaction in Games and New Media," in *From Pac-Man to Pop Music: Interactive Audio in Games and New Media*, ed. Karen Collins (Burlington, VT: Ashgate, 2008), 104.

42. Bryan W. C. Chung, *Multimedia Programming with Pure Data* (Birmingham: Packt, 2013); Eric Lyon, *Designing Audio Objects for Max/MSP and Pd* (Middleton, WI: A-R Editions, 2012).

43. Bessell, "Blurring Boundaries."

44. Rémi, "The Maturity of Visual Programming," *Craft AI*, September 29, 2015, http://www.craft.ai/blog/the-maturity-of-visual-programming/.

45. Jason Kincaid, "RjDj Generates an Awesome, Trippy Soundtrack for Your Life," *TechCrunch*, October 13, 2008, http://techcrunch.com/2008/10/13/rjdj-generates-an-awesome-trippy-soundtrack-for-your-life/.

46. Quoted in "Monu/Mental," Unity3D, January 23, 2014, https://unity3d.com/showcase/case-stories/monument-valley. Emphasis added.

47. Jon Brodkin, "How Unity3D Became a Game-Development Beast," *Dice*, June 3, 2013, https://insights.dice.com/2013/06/03/how-unity3d-become-a-game-development-beast/.

48. Quoted in "Rural Legend," Unity3D, April 25, 2014, https://unity3d.com/showcase/case-stories/endless-legend. Emphasis added.

49. Artur Ganszyniec, "In Praise of Slow Games," *Gamasutra*, June 24, 2019, https://www.gamasutra.com/blogs/ArturGanszyniec/20190624/345350/In_Praise_of_Slow_Games.php.

50. Benjamin, *Race after Technology*, 180.

51. Anna Everett and S. Craig Watkins, "The Power of Play: The Portrayal and Performance of Race in Video Games," in *The Ecology of Games: Connecting Youth, Games, and Learning*, ed. Katie Salen (Cambridge, MA: MIT Press, 2008), 58.

52. Maisa Imamović, "How to Nothing," Institute of Network Cultures, January 14, 2019, https://networkcultures.org/longform/2019/01/14/how-to-nothing/.

Chapter 7

1. "IBM Watson Music," IBM, n.d., https://www.ibm.com/watson/music/uk-en/ .
2. "How IBM Watson Inspired Alex Da Kid's New Song 'Not Easy,'" *Business Insider*, October 25, 2016, https://www.businessinsider.com/sc/ibm-watson-helps-cre ate-alex-da-kid-song-2016-10.
3. "Innovating for Writers and Artists," Spotify, July 12, 2017, https://artists.spotify. com/blog/innovating-for-writers-and-artists.
4. "IBM Watson Music."
5. Shahan Nercessian, "iZotope and Assistive Audio Technology," iZotope, July 19, 2018, https://www.izotope.com/en/learn/izotope-and-assistive-audio-technology.html.
6. Alexander Weheliye, "Desiring Machines in Black Popular Music," in *The Sound Studies Reader*, ed. Jonathan Sterne (New York: Routledge, 2012), 511–519.
7. Scott Wilson, "The 14 Drum Machines That Shaped Modern Music," *FACTmag*, September 22, 2016, https://www.factmag.com/2016/09/22/the-14-drum-machines- that-shaped-modern-music/.
8. "BFD3: Evolved Acoustic Drum Software," FXpansion, n.d., https://www.fxpansion. com/products/bfd3/.
9. Anne Danielsen and Ragnhild Brøvig-Hanssen, *Digital Signatures: The Impact of Digitization on Popular Music Sound* (Cambridge, MA: MIT Press, 2016).
10. "Meet Liquid Rhythm," WaveDNA, https://www.wavedna.com/liquid-rhythm/.
11. "Rhythmiq," Accusonus, https://accusonus.com/products/rhythmiq.
12. Google, "Magenta Studio (v1.0)," Magenta, n.d., https://magenta.tensorflow.org/ studio.
13. "Flow Machines: AI Assisted Music," Sony, n.d., https://www.flow-machines.com.
14. Christine McLeavey Payne, "MuseNet," OpenAI, April 25, 2019, https://openai.com/ blog/musenet/.
15. "Online Audio Mastering," LANDR, https://www.landr.com/en/online-audio- mastering.
16. Amper Music, n.d., https://www.ampermusic.com.
17. "Why Do We Exist?" CloudBounce, https://www.cloudbounce.com/about#why.
18. Josh Dzieza, "How Hard Will the Robots Make Us Work?" *The Verge*, February 27, 2020, https://www.theverge.com/2020/2/27/21155254/automation-robots-unemp loyment-jobs-vs-human-google-amazon.
19. Dani Deahl, "We've Been Warned about AI and Music for over 50 Years, but No One's Prepared," *The Verge*, April 17, 2019, https://www.theverge.com/2019/4/17/18299 563/ai-algorithm-music-law-copyright-human.
20. Quoted in Bell, *Dawn of the DAW*, 142–143.
21. Richard Leppert, "Music 'Pushed to the Edge of Existence' (Adorno, Listening, and the Question of Hope)," *Cultural Critique* 60 (2005): 116.
22. Jacques Attali, *Noise: The Political Economy of Music* (Minneapolis: University of Minnesota Press, 1985), 6. See also Robin James, *The Sonic Episteme: Acoustic Resonance, Neoliberalism, and Biopolitics* (Durham, NC: Duke University Press, 2019).

23. Robert W. Taylor, "Hyper-Compression in Music Production: Agency, Structure and the Myth That 'Louder Is Better,'" *Journal on the Art of Record Production* 11 (2017): n.p., https://www.arpjournal.com/asarpwp/hyper-compression-in-music-production-agency-structure-and-the-myth-that-louder-is-better/.

24. Future Music, "What Is Machine Learning, and What Does It Mean for Music?" *Music Radar*, August 30, 2019, https://www.musicradar.com/news/what-is-machine-learning-and-what-does-it-mean-for-music.

25. Kariann Goldschmitt and Nick Seaver, "Shaping the Stream: Techniques and Troubles of Algorithmic Recommendation," in *Cambridge Companion to Music and Digital Culture*, ed. Nicholas Cook et al. (Cambridge: Cambridge University Press, 2019); Eric Drott, "Why the Next Song Matters: Streaming, Recommendation, Scarcity," *Twentieth-Century Music* 15, no. 3 (2018): 325–357; Patrick Vonderau, "The Politics of Content Aggregation," *Television & New Media* 16, no. 8 (2015): 717–733.

26. Marinos Koutsomichalis, "From Music to Big Music: Listening in the Age of Big Data," *Leonardo Music Journal* 26 (2016): 26.

27. Hal Foster, *Design and Crime (and Other Diatribes)* (London: Verso, 2002), 17.

28. Sara M. Watson, "Data Doppelgängers and the Uncanny Valley of Personalization," *The Atlantic*, June 16, 2014, \ https://www.theatlantic.com/technology/archive/2014/06/data-doppelgangers-and-the-uncanny-valley-of-personalization/372780/. For other critiques of big data, see "Persuasion and the Other Thing: A Critique of Big Data Methodologies in Politics," *Ethnography Matters*, May 24, 2017, https://ethnographymatters.net/blog/2017/05/24/persuasion-and-the-other-thing-a-critique-of-big-data-methodologies-in-politics/; Cathy O'Neill, *Weapons of Mass Destruction: How Big Data Increases Inequality and Threatens Democracy* (New York: Crown, 2016).

29. Valerie Ng and Jon Tilliss, "Balancing Aesthetics and Usability in Medical User Interface Design: 9 Key Trends," *Emergo*, July 8, 2018, https://www.emergobyul.com/blog/2018/07/balancing-aesthetics-and-usability-medical-user-interface-design-9-key-trends.

30. Karin Bijsterveld, "Listening to Machines: Industrial Noise, Hearing Loss and the Cultural Meaning of Sound," in *The Sound Studies Reader*, ed. Jonathan Sterne (Durham, NC: Duke University Press, 2012): 152–167.

31. Alan Williams, "Putting It on Display: The Impact of Visual Information on Control Room Dynamics," *Journal on the Art of Record Production* 6 (2012): n.p., https://www.arpjournal.com/asarpwp/putting-it-on-display-the-impact-of-visual-information-on-control-room-dynamics/.

32. Btihaj Ajana, "Digital Health and the Biopolitics of the Quantified Self," *Digital Health* 3 (2017): 1–18.

33. Catherine Provenzano, "Making Voices: The Gendering of Pitch Correction and the Auto-Tune Effect in Contemporary Pop Music," *Journal of Popular Music Studies* 31, no. 2 (2019): 63–84.

34. "Insight 2," iZotope, n.d., https://www.izotope.com/en/products/insight.html.

35. Emily Thompson writes about how the control of noise was a central goal in the modernization of America throughout the twentieth century. *The Soundscape of*

Modernity: Architectural Acoustics and the Culture of Listening in America, 1900–1933 (Cambridge, MA: The MIT Press, 2002).

36. Sam Machkovech, "SoundCloud's Free 'Auto-Mastering' Audio Tool Is More of an Auto-Turd," *Ars Technica*, May 29, 2016, https://arstechnica.com/gadgets/2016/05/soundclouds-free-auto-mastering-audio-tool-is-more-of-an-auto-turd/.

37. Rob Stewart, "AI-Based Mastering, Human Audio Mastering and the Future," *JustMastering*, May 29, 2016, https://www.justmastering.com/article-landr-audio-mastering.php.

38. Bobby Owsinski, "Is Mastering an Endangered Audio Job?" Music Production Blog, July 10, 2019, https://bobbyowsinskiblog.com/2019/07/10/mastering-endangered/.

39. Brecht De Man and Joshua D. Reiss, "A Semantic Approach to Autonomous Mixing," *Journal on the Art of Record Production* 8 (2013): n.p., https://www.arpjournal.com/asarpwp/a-semantic-approach-to-autonomous-mixing/; Mike Senior, *Mixing Secrets* (New York: Routledge, 2012); Alex Case, *Sound FX: Unlocking the Creative Potential of Recording Studio Effects* (New York: Routledge, 2012); Alex Case, *Mix Smart: Professional Techniques for the Home Studio* (New York: Routledge, 2011).

40. Jordan Kisner, "The Dark Art of Mastering Music," *Pitchfork*, May 19, 2016, https://pitchfork.com/features/article/9894-the-dark-art-of-mastering-music/.

41. Quoted in ibid.

42. Alex Holmes, "Review: iZotope Ozone 9," *MusicTech*, October 29, 2019, https://www.musictech.net/reviews/plug-ins/izotope-ozone-9/.

43. De Man and Reiss, "A Semantic Approach."

44. Bobby Owsinski, *The Mastering Engineer's Handbook*, 4th ed. (Burbank, CA: Bobby Owsinski Media Group, 2017), 23.

45. Anahid Kassabian, *Ubiquitous Listening: Affect, Attention, and Distributed Subjectivity* (Berkeley: University of California Press, 2013).

46. Pauline Oliveros, *Deep Listening: A Composer's Sound Practice* (Lincoln, NE: iUniverse, 2005).

Conclusion

1. Zachary B. Wolf, "This Is What Coronavirus Capitalism Looks Like," CNN, April 28, 2020, https://www.cnn.com/2020/04/28/politics/what-matters-april-27/index.html; Naomi Klein, "Coronavirus Capitalism—and How to Beat It," *The Intercept*, March 16, 2020, https://theintercept.com/2020/03/16/coronavirus-capitalism/.

2. Mya Guarnieri, "Stop Saying 'We're All in This Together.' You Have Money. It's Not the Same," *Washington Post*, April 18, 2020, https://www.washingtonpost.com/outlook/2020/04/18/coronavirus-retail-jobs-inequality/.

3. Endlesss, "Products," n.d., https://endlesss.fm/products.

4. "Summer Seasonal Gaming Guide," *Polygon*, n.d., https://www.polygon.com/2020/6/1/21273160/summer-seasonal-gaming-guide.

5. Imad Khan, "Why Animal Crossing Is the Game for the Coronavirus Moment," *New York Times*, April 7, 2020, https://www.nytimes.com/2020/04/07/arts/animal-crossing-covid-coronavirus-popularity-millennials.html; Anne Wallace, "'Animal Crossing: New Horizons' Is Exactly What We Need Right Now," *Deseret News*, March 23, 2020, https://www.deseret.com/entertainment/2020/3/23/21191101/animal-crossing-new-horizons-nintendo-coronavirus-social-distancing-quarantine-covid-19; Annie Jennemann, "'Animal Crossing: New Horizons' Is the Perfect Game for Self-Isolating," *Vox*, March 26, 2020, https://www.voxmagazine.com/arts/animal-crossing-new-horizons-is-the-perfect-game-for-self-isolating/article_84d78f9a-6ee7-11ea-a7c0-4f0900a6b9a6.html; Jan Cortes, "The Reason Why 'Animal Crossing' Is the Perfect Pandemic Pastime," *Medical Daily*, April 8, 2020, https://www.medicaldaily.com/reason-why-animal-crossing-perfect-pandemic-pastime-451602.

6. Gita Jackson, "'Animal Crossing: New Horizons' Is Not the Game We All Need Right Now," *Vice*, April 14, 2020, https://www.vice.com/en_us/article/jgezwk/animal-crossing-new-horizons-is-not-the-game-we-all-need-right-now.

7. "Oculus Quest," Oculus, n.d., https://www.oculus.com/quest/?locale=en_US.

8. Benjamin, *Race after Technology*, 172.

9. "Electronauts," Survios, n.d., https://survios.com/electronauts/.

10. Chun, *Programmed Visions*, 59.

11. Ibid., 67.

12. Ibid., 17.

13. Ibid., 90.

14. Christian Ulrik Andersen and Søren Bro Pold, *The Metainterface: The Art of Platforms, Cities, and Clouds* (Cambridge, MA: MIT Press, 2018), 138.

15. Brian Massumi, *Parables for the Virtual: Movement, Affect, Sensation* (Durham, NC: Duke University Press, 2002), 6.

16. Quoted in David Parisi, *Archaeologies of Touch: Interfacing with Haptics from Electricity to Computing* (Minneapolis: University of Minnesota Press, 2018), 3.

17. Judith Butler, *Giving an Account of Oneself* (New York: Fordham University Press, 2005), 20.

18. Adriana Cavarero, *Relating Narratives: Storytelling and Selfhood* (London: Psychology Press, 2000), 87.

19. Erin McCarthy, *Ethics Embodied: Rethinking Selfhood through Continental, Japanese, and Feminist Philosophies* (Lanham, MD: Rowman & Littlefield, 2010), 25.

20. Cavarero, *Relating Narratives*, 84.

21. Butler, *Giving an Account*, 136.

22. Nusselder, *Interface Fantasy*, 9.

23. Peter Horsfield, "Continuities and Discontinuities in Ethical Reflections on Digital Virtual Reality," *Journal of Mass Media Ethics* 34 nos. 3–4 (2003): 159.

24. See Spencer Ackerman and Noah Shachtman, "Almost 1 in 3 U.S. Warplanes Is a Robot," *Wired*, January 1, 2012, http://www.wired.com/dangerroom/2012/01/drone-report/; Katie Drummond, "Army's Virtual Reality Plan: A Digital Doppelganger for Every Soldier," *Wired*, January 18, 2012, http://www.wired.com/dangerroom/2012/01/army-virtual-reality/.

25. McGonigal, *Reality Is Broken*, 24.

26. Tom Boellstorff, *Coming of Age in Second Life: An Anthropologist Explores the Virtually Human* (Princeton, NJ: Princeton University Press, 2015), 206.

27. Robert Romanyshyn suggests that our experience of technology has become a dream "of domination, mastery, and control of nature," in *Technology as Symptom and Dream* (New York: Routledge, 1989), 211.

28. Stephen Stockwell and Adam Muir, "The Military-Entertainment Complex: A New Facet of Information Warfare," *Fibreculture Journal* 1 (2003), http://one.fibreculture journal.org/fcj-004-the-military-entertainment-complex-a-new-facet-of-informat ion-warfare.

29. Ray Kurzweil, quoted in Kate Lunau, "Google's Ray Kurzweil on the Quest to Live Forever," *Maclean's,* October 14, 2013, https://www.macleans.ca/society/life/how-nanobots-will-help-the-immune-system-and-why-well-be-much-smarter-thanks-to-machines-2/.

30. Clifford Geertz, "Thick Description: Toward an Interpretative Theory of Culture," in *The Interpretation of Cultures* (New York: Basic Books, 1973), 3–36.

31. McPherson, "Why Are the Digital Humanities So White?"; Tara Rodgers, "Towards a Feminist Historiography of Electronic Music," in *The Sound Studies Reader*, ed. Jonathan Sterne, 475–489 (New York: Routledge, 2012); Mara Mills, "Do Signals Have Politics? Inscribing Abilities in Cochlear Implants," in *The Oxford Handbook of Sound Studies*, ed. Trevor Pinch and Karin Bijsterveld, 320–346 (New York: Oxford University Press, 2012).

32. Susan Wojcicki, "Closing the Tech Industry Gender Gap," *Huffington Post*, January 27, 2016, http://www.huffingtonpost.com/susan-wojcicki/tech-industry-gender-gap_b_9089472.html; Lauren Gilmore, "The Gender Gap in Technology Can Byte Me," *The Next Web*, December 1, 2015, http://thenextweb.com/insider/2015/12/01/the-gender-gap-in-technology-can-byte-me/#gref; Catherine Bracy, "Closing the Tech Industry's Gender Gap Requires Better Data," NPR, June 25, 2013, http://www.npr.org/sections/alltechconsidered/2013/06/24/195144754/closing-the-tech-industry-s-gender-gap-requires-better-data.

33. Ken Olsen, quoted in Jonathan Gatlin, *Bill Gates: The Path to the Future* (New York: HarperCollins, 1999), 39.

34. Rob Beschizza, "The 15 Dumbest Apple Predictions of All Time," *Wired*, November 1, 2007, http://www.wired.com/2007/11/analysts-dont-k/.

Bibliography

Abbate, Carolyn. "Music—Drastic or Gnostic." *Critical Inquiry* 30, no. 3 (2004): 505–536.

Ableton. "Ableton and Cycling '74 Form a New Partnership." June 7, 2017. https://www.ableton.com/en/blog/ableton-cycling-74-new-partnership/.

Ableton. "Ander's Incredible Custom Controller for Live." Ableton, July 13, 2012. https://www.ableton.com/en/blog/anders-incredible-custom-controller-live/.

Ableton. "Artist Quotes." Ableton, n.d. https://www.ableton.com/en/pages/artists/artist_quotes/.

Ableton. "Machinedrum: Sacred Frequencies." September 25, 2012. https://www.ableton.com/en/blog/machinedrum/.

Ableton. "What Is This Book?" http://makingmusic.ableton.com/about.

"Ableton Live Audio & Loop Sequencing Software for Mac and Windows." *Beat Mode*, n.d. http://www.beatmode.com/historical/ableton-live/.

Abtan, Freida. "Where Is She? Finding the Women in Electronic Music Culture." *Contemporary Music Review* 35, no. 1 (2016): 53–60.

Ackerman, Spencer, and Noah Shachtman. "Almost 1 in 3 U.S. Warplanes Is a Robot." *Wired*, January 9, 2012. http://www.wired.com/dangerroom/2012/01/drone-report/.

Adorno, Theodor W. *The Culture Industry: Selected Essays on Mass Culture*. New York: Routledge, 1991.

Ajana, Btihaj. "Digital Health and the Biopolitics of the Quantified Self." *Digital Health* 3 (2017): 1–18.

Akrich, Madeleine. "The De-Scription of Technical Objects." In *Shaping Technology/Building Society: Studies in Sociotechnical Change*, edited by Wiebe E. Bijker and John Law, 205–224. Cambridge, MA: MIT Press, 1992.

Allan, Roy A. *A History of the Personal Computer: The People and the Technology*. London, ON: Allan, 2001. https://archive.org/details/A_History_of_the_Personal_Computer.

Amper Music, n.d. https://www.ampermusic.com.

Andersen, Christian Ulrik, and Søren Bro Pold. *Interface Criticism: Aesthetics beyond Buttons*. Aarhus: Aarhus University Press, 2011.

Andersen, Christian Ulrik, and Søren Bro Pold. *The Metainterface: The Art of Platforms, Cities, and Clouds*. Cambridge, MA: MIT Press, 2018.

"Angélica Negrón: Toys Noise." Ableton, January 13, 2012. https://www.ableton.com/en/blog/angelica-negron/.

Antidote, Sam. "Top 5 Reference Tracks for Producers." *Antidote Audio*, October 19, 2017. https://www.antidoteaudio.com/blog-reference-songs.

Apple. "Creativity Apps: iOS." N.d. http://www.apple.com/si/creativity-apps/ios/.

Apple. "Designing for iOS." In *Human Interface Guidelines*. Cupertino, CA: Apple, 2014. https://developer.apple.com/library/iOS/documentation/userexperience/conceptual/mobilehig/.

Apple. "Gestures." In *Human Interface Guidelines*, https://developer.apple.com/design/human-interface-guidelines/ios/user-interaction/gestures/.

Apple. "Switch Apps on Your iPhone, iPad, or iPod Touch." December 1, 2020. https://support.apple.com/en-us/HT202070.

Apple. "Use Multi-Touch Gestures on Your Mac." N.d. https://support.apple.com/en-us/HT204895.

Armstrong, Victoria. "Hard Bargaining on the Hard Drive: Gender Bias in the Music Technology Classroom." *Gender and Education* 20, no. 4 (2008): 375–386.

Armstrong, Victoria. *Technology and the Gendering of Music Education.* Farnham, UK: Ashgate, 2011.

Arzoomanian, Rich. "A Complete History of Mainframe Computing." *Tom's Hardware*, June 26, 2009. https://www.tomshardware.com/picturestory/508-mainframe-computer-history.html.

Atherton, Kelsey. "Microsoft Is Training AI in Minecraft." *Popular Science*, March 14, 2016. http://www.popsci.com/microsoft-is-training-ai-in-minecraft.

Attali, Jacques. *Noise: The Political Economy of Music.* Minneapolis: University of Minnesota Press, 1985.

Auslander, Philip. *Liveness: Performance in a Mediatized Culture.* New York: Routledge, 2008.

Barbera, Allie. "Professor, Songwriter, Producer, Musician, Music Technology Consultant, and Now Soon-to-Be Mama: Erin Barra-Jean." *JSJEvents*, October 23, 2017. https://www.jsjeventsboston.com/blog/2017/9/25/getting-to-know-erin-barra-founder-of-beatz-by-girls.

Bardzell, Shaowen. "Feminist HCI: Taking Stock and Outlining an Agenda for Design." In *Proceedings of the SIGCHI Conference on Human Factors in Computing Systems*, 1301–1310. New York: Association for Computing Machinery, 2010.

Bardzell, Shaowen. "Through the 'Cracks and Fissures' in the Smart Home to Ubiquitous Utopia." In *Critical Theory and Interaction Design*, edited by Jeffrey Bardzell, Shaowen Bardzell, and Mark Blythe, 751–779. Cambridge, MA: MIT Press, 2018.

Barr, Pippin, James Noble, and Robert Biddle. "Video Game Values: Human-Computer Interaction and Games." *Interacting with Computers* 19 (2007): 180–195.

Barrett, Brian. "End of an Era: Panasonic Kills Off Technics Turntables." *Gizmodo*, October 28, 2010. http://gizmodo.com/5675818/end-of-an-era-panasonic-kills-off-technics-turntables.

Bates, Eliot. "The Social Life of Musical Instruments." *Ethnomusicology* 54, no. 3 (2012): 363–395.

Bates, Eliot. "What Studios Do." *Journal on the Art of Record Production* 7 (2012): n.p. http://arpjournal.com/what-studios-do/.

Baudouin, Oliver. "A Reconstruction of *Stria.*" *Computer Music Journal* 31, no. 3 (2007): 75–81.

Bauman, Valerie. "78% of Women in the Music Industry Say They're Treated Differently at Work Because of Their Gender, Study Says." *Daily Mail*, March 12, 2019. https://www.dailymail.co.uk/news/article-6801023/78-Women-music-industry-say-theyre-treated-differently-work-gender.html.

Bayton, Mavis. "Women and the Electric Guitar." In *Sexing the Groove: Popular Music and Gender*, edited by Sheila Whiteley, 37–49. New York: Routledge, 1997.

Beats By Girls. N.d. https://beatsbygirlz.org.

Bell, Adam Patrick. "DAW Democracy? The Dearth of Diversity in 'Playing the Studio.'" *Journal of Music, Technology & Education* 8, no. 2 (2015): 129–146.

Bell, Adam Patrick. *Dawn of the DAW: The Studio as Musical Instrument.* New York: Oxford University Press, 2018.

Bell, Adam Patrick, Ethan Hein, and Jarrod Ratcliffe. "Beyond Skeuomorphism: The Evolution of Music Production Software User Interface Metaphors." *Journal on the Art of Record Production* 9 (2015): n.p. https://www.arpjournal.com/asarpwp/beyond-skeuomorphism-the-evolution-of-music-production-software-user-interface-metaph ors-2/.

Belshaw, Doug. *The Essential Elements of Digital Literacies*. Self-published, 2014. https:// dougbelshaw.com/essential-elements-book.pdf.

Bench, Harmony. "Gestural Choreographies: Embodied Disciplines and Digital Media." In *The Oxford Handbook of Mobile Music Studies*, Vol. 2, edited by Sumanth Gopinath and Jason Stanyek, 238–256. New York: Oxford University Press, 2014.

Benet-Weiser, Sarah. *Empowered: Popular Feminism and Popular Misogyny*. Durham, NC: Duke University Press, 2018.

"Benga Makes a Grime Beat on PlayStation Music 2000—BBC 2003." GetDarker, December 5, 2014. https://www.youtube.com/watch?v=AZHnaSIZP0Y.

Benjamin, Ruha. *Race after Technology: Abolitionist Tools for the New Jim Code*. Cambridge: Polity, 2019.

Bergh, Arild, and Tia DeNora. "From Wind-Up to iPod: Techno-Cultures of Listening." In *The Cambridge Companion to Recorded Music*, edited by Nicholas Cook, Eric Clarke, Daniel Leech-Wilkinson, and John Rink, 102–115. Cambridge, UK: Cambridge University Press, 2009.

Bergstrom, Ilias, and R. Beau Lotto. "Code Bending: A New Creative Coding Practice." *Leonardo* 48, no. 1 (2015): 25–31.

Bernstein, David W. *The San Francisco Tape Music Center: 1960s Counterculture and the Avant-Garde*. Berkeley: University of California Press, 2008.

Berry, David. *The Philosophy of Software: Code and Mediation in the Digital Age*. New York: Palgrave Macmillan, 2011.

Berry, David M., and Michael Dieter, eds. *Postdigital Aesthetics: Art, Computation and Design*. New York: Palgrave Macmillan, 2015.

Beschizza, Rob. "The 15 Dumbest Apple Predictions of All Time." *Wired*, November 1, 2007. http://www.wired.com/2007/11/analysts-dont-k/.

Bessell, David. "Blurring Boundaries: Trends and Implications in Audio Production Software Developments." In *The Oxford Handbook of Interactive Audio*, edited by Karen Collins, Bill Kapralos, and Holly Tessler, 405–418. New York: Oxford University Press, 2013.

"BFD3: Evolved Acoustic Drum Software." FXpansion, n.d. https://www.fxpansion.com/ products/bfd3/.

Bijker, Wiebe E., Thomas P. Hughes, and Trevor Pinch. *The Social Construction of Technological Systems*. Cambridge: MIT Press, 2012.

Bijsterveld, Karin. "Listening to Machines: Industrial Noise, Hearing Loss and the Cultural Meaning of Sound." In *The Sound Studies Reader*, edited by Jonathan Sterne, 152–167. Durham, NC: Duke University Press, 2012.

Bissell, Tom. *Extra Lives: Why Video Games Matter*. New York: Pantheon Books, 2010.

Blythe, Mark. "After Critical Design." In *Critical Theory and Interaction Design*, edited by Jeffrey Bardzell, Shaowen Bardzell, and Mark Blythe. Cambridge, MA: MIT Press, 2018.

Boellstorff, Tom. *Coming of Age in Second Life: An Anthropologist Explores the Virtually Human*. Princeton, NJ: Princeton University Press, 2015.

Bogost, Ian. *Persuasive Games: The Expressive Power of Videogames*. Cambridge and London: The MIT Press, 2007.

Bogost, Ian. "Persuasive Games: The Proceduralist Style." *Gamasutra*, January 21, 2009. http://www.gamasutra.com/view/feature/132302/persuasive_games_the_.php.

Bogost, Ian. *Unit Operations: An Approach to Videogame Criticism*. Cambridge, MA: MIT Press, 2006.

Bollen, Jonathan. "Queer Kinesthesia: Performativity on the Dance Floor." In *Dancing Desires: Choreographing Sexualities On and Off the Stage*, edited by Jane C. Desmond, 285–314. Madison: University of Wisconsin Press, 2001.

Bolter, Jay David, and Richard Grusin. *Remediation: Understanding New Media*. Cambridge, MA: MIT Press, 2000.

Borison, Rebecca. "The 15 Highest-Grossing iPhone and iPad Games." *Business Insider*, May 20, 2014. http://www.businessinsider.com/highest-grossing-iphone-and-ipad-games-2014-5?op=1.

Born, Georgina. "On Musical Mediation: Ontology, Technology, and Creativity." *Twentieth-Century Music* 2, no. 1 (2005): 7–36.

Born, Georgina. *Rationalizing Culture: IRCAM, Boulez, and the Institutionalization of the Musical Avant-Garde*. Berkeley: University of California Press, 1995.

Born, Georgina. "Recording: From Reproduction to Representation to Remediation." In *The Cambridge Companion to Recorded Music*, edited by Nicholas Cook, Eric Clarke, Daniel Leech-Wilkinson, and John Rink, 286–304. Cambridge: Cambridge University Press, 2009.

Born, Georgina, and Kyle Devine. "Introduction: Gender, Creativity and Education in Digital Musics and Sound Art." *Contemporary Music Review* 35, no. 1 (2016): 6–7.

Borzyskowski, George. "The Hacker Demo Scene and It's [*sic*] Cultural Artefacts." http://citeseerx.ist.psu.edu/viewdoc/download?doi=10.1.1.130.4968&rep=rep1&type=pdf.

Böttcher, Niels, Héctor P. Martinez, and Stefania Serafin. "Procedural Audio in Computer Games Using Motion Controllers: An Evaluation on the Effect and Perception." *International Journal of Computer Games Technology* (2013).

Bracy, Catherine. "Closing the Tech Industry's Gender Gap Requires Better Data." NPR, June 25, 2013. http://www.npr.org/sections/alltechconsidered/2013/06/24/195144754/closing-the-tech-industry-s-gender-gap-requires-better-data.

Brand, Stewart. "Spacewar." *Rolling Stone*, December 7, 1972. http://www.wheels.org/spacewar/stone/rolling_stone.html.

Brodkin, Jon. "How Unity3D Became a Game-Development Beast." *Dice*, June 3, 2013. http://news.dice.com/2013/06/03/how-unity3d-become-a-game-development-beast/.

Bromley, Hank, and Michael W. Apple, eds. *Education/Technology/Power: Educational Computing as Social Practice*. Albany: State University of New York Press, 1998.

Brown, Peter. *Information Architecture with XML*. Hoboken, NJ: John Wiley, 2003.

Burgess, Jean. "The iPhone Moment, the Apple Brand, and the Creative Consumer: From 'Hackability and Usability' to Cultural Generativity." In *Studying Mobile Media: Cultural Technologies, Mobile Communication, and the iPhone*, edited by Larissa Hjorth, Jean Burgess, and Ingrid Richardson, 28–42. New York: Routledge, 2012.

Burgess, Jean. "Vernacular Creativity and New Media." PhD diss., University of Queensland, 2007.

Butler, Chris. *Henri Lefebvre: Spatial Politics, Everyday Life and the Right to the City*. New York: Routledge, 2012.

Butler, Judith. *Giving an Account of Oneself*. New York: Fordham University Press, 2005.

Butler, Mark. *Playing with Something That Runs: Technology, Improvisation, and Composition in DJ and Laptop Performance*. New York: Oxford University Press, 2014.

Butler, Shane. *The Ancient Phonograph*. Cambridge, MA: MIT Press, 2015.

Cage, John. *Silence: Lectures and Writings*. Hanover, NH: Wesleyan University Press, 1961.

Callon, Michel, and Bruno Latour. "Unscrewing the Big Leviathan: How Actors Macrostructure Reality and How Sociologists Help Them to Do So." In *Advances in Social Theory and Methodology: Toward an Integration of Micro- and Macro-Sociologies*, edited by K. D. Knorr-Cetina and A. V. Cicourel, 277–303. Boston: Routledge and Kegan Paul, 1981.

Cariani, Peter. "Emergence and Artificial Life." In *Artificial Life II*, edited by Christopher G. Langton, Charles Taylor, J. Doyne Farmer, and Steen Rasmussen, 775–798. New York: Addison-Wesley, 1991.

Carlson, Glen. "Emergent Aesthetics." http://www.glencarlson.com/emergent/Carlson_ EmergentAesthetics.pdf.

Carr, Nicholas. *The Shallows: What the Internet Is Doing to Our Brains*. New York: W. W. Norton, 2011.

Cascella, Daniela. "Sound Objects: Pierre Schaeffer's 'In Search of a Concrete Music.'" *Los Angeles Review of Books*, April 3, 2013. http://lareviewofbooks.org/review/sound-obje cts-pierre-schaeffers-in-search-of-a-concrete-music.

Cascone, Kim. "The Aesthetics of Failure: 'Post-Digital' Tendencies in Contemporary Computer Music." *Computer Music Journal* 24, no. 4 (2000): 12–18.

Case, Alex. *Mix Smart: Professional Techniques for the Home Studio*. New York: Routledge, 2011.

Case, Alex. *Sound FX: Unlocking the Creative Potential of Recording Studio Effects*. New York: Routledge, 2012.

Cavarero, Adriana. *Relating Narratives: Storytelling and Selfhood*. London: Psychology Press, 2000.

Ceruzzi, Paul E. *Computing: A Concise History*. Cambridge, MA: MIT Press, 2012.

Cheng, William. *Sound Play: Video Games and the Musical Imagination*. Oxford: Oxford University Press, 2014.

Chun, Wendy Hui Kyong. *Programmed Visions: Software and Memory*. Cambridge: MIT Press, 2011.

Chun, Wendy Hui Kyong. *Updating to Remain the Same: Habitual New Media*. Cambridge, MA: MIT Press, 2016.

Chung, Bryan W. C. *Multimedia Programming with Pure Data*. Birmingham: Packt, 2013.

Citron, Marcia J. *Gender and the Musical Canon*. Cambridge: Cambridge University Press, 1993.

Clayton, Jace. *Uproot: Travels in 21st-Century Music and Digital Culture*. New York: Farrar, Straus and Giroux, 2016.

Clegg, S. "Theorising the Machine: Gender, Education and Computing." *Gender and Education* 13, no. 3 (2001): 307–324.

"Clint Sand: Electronic Polymath." Ableton, January 20, 2011. https://www.ableton.com/ en/pages/artists/clint_sand/.

Code.org. "Hour of Code." N.d. http://hourofcode.com/us.

Colley, A., and C. Comber. "Age and Gender Differences in Computer Use and Attitudes among Secondary School Students: What Has Changed?" *Educational Research* 45, no. 2 (2003): 155–165.

Collins, Karen, ed. *From Pac-Man to Pop Music: Interactive Audio in Games and New Media*. Burlington, VT: Ashgate, 2008.

Collins, Karen. "An Introduction to Procedural Music in Video Games." *Contemporary Music Review* 28, no. 1 (2009): 5–15.

Collins, Karen. *Playing with Sound: A Theory of Interacting with Sound and Music in Video Games*. Cambridge, MA: MIT Press, 2013.

Collins, Karen, Bill Kapralos, and Holly Tessler, eds. *The Oxford Handbook of Interactive Audio*. New York: Oxford University Press, 2013.

Collins, Nick. "Live Coding of Consequence." *Leonardo* 44, no. 3 (2011): 207–211.

Cook, Daniel. "What Are Game Mechanics?" *Lostgarden*, October 23, 2006. http://www. lostgarden.com/2006/10/what-are-game-mechanics.html.

Cortes, Jan. "The Reason Why 'Animal Crossing' Is the Perfect Pandemic Pastime." *Medical Daily*, April 8, 2020. https://www.medicaldaily.com/reason-why-animal-cross ing-perfect-pandemic-pastime-451602.

Cowen, Tyler. "The Next Silicon Valley Could Be . . . Los Angeles?" Bloomberg, June 26, 2019.   https://www.bloomberg.com/opinion/articles/2019-06-26/los-angeles-could- be-the-next-silicon-valley.

Cox, Geoff, and Alex Mclean. *Speaking Code: Coding as Aesthetic and Political Expression*. Cambridge, MA: MIT Press, 2012.

Crowther, Robert D., and Kevin Durkin. "Sex- and Age-Related Differences in the Musical Behavior, Interests and Attitudes towards Music of 232 Secondary School Students." *Educational Studies* 8, no. 2 (1982): 131–139.

"Daedelus, Live at Ground Kontrol: Far from Home #9." Intothewoods.tv, July 9, 2012. https://www.youtube.com/watch?v=QY3mHBj_dhs.

Danielsen, Anne, and Ragnhild Brøvig-Hanssen. *Digital Signatures: The Impact of Digitization on Popular Music Sound*. Cambridge, MA: MIT Press, 2016.

Daub, Adrian. *What Tech Calls Thinking: An Inquiry into the Intellectual Bedrock of Silicon Valley*. New York: Farrar, Straus and Giroux, 2020.

Deadmau5. "We All Hit Play." *United We Fail*, 2013. http://deadmau5.tumblr.com/post/ 25690507284/we-all-hit-play.

Deahl, Dani. "We've Been Warned about AI and Music for over 50 Years, but No One's Prepared." *The Verge*, April 17, 2019. https://www.theverge.com/2019/4/17/18299563/ ai-algorithm-music-law-copyright-human.

"Decap and Brady Watt on Push & Working with Ski Beatz." Ableton, April 16, 2014. https://www.ableton.com/en/blog/decap-and-brady-watt-push-working-ski-beatz/.

Décary-Hétu, David. "Police Operations 3.0: On the Impact and Policy Implications of Police Operations on the Warez Scene." *Policy and Internet* 6, no. 3 (2014): 315–340.

Dehouk, Rémi. "The Maturity of Visual Programming." *craft ai*, September 29, 2015. http://www.craft.ai/blog/the-maturity-of-visual-programming/.

Delzell, Judith K., and David A. Leppla. "Gender Association of Musical Instruments and Preferences of Fourth-Grade Students for Selected Instruments." *Journal of Research in Music Education* 40, no. 2 (1992): 93–103.

De Man, Brecht, and Joshua D. Reiss. "A Semantic Approach to Autonomous Mixing." *Journal on the Art of Record Production* 8 (2013): n.p. https://www.arpjournal.com/asar pwp/a-semantic-approach-to-autonomous-mixing/.

D'Errico, Mike. "Going Hard: Bassweight, Sonic Warfare, and the 'Brostep' Aesthetic." *Sounding Out!*, January 23, 2014. https://soundstudiesblog.com/2014/01/23/going- hard-bassweight-sonic-warfare-the-brostep-aesthetic/.

Dery, Mark. "Black to the Future: Interviews with Samuel R. Delany, Greg Tate, and Tricia Rose." In *Flame Wars: The Discourse of Cyberculture*, edited by Mark Dery, 179–222. Durham, NC: Duke University Press, 1994.

DeSantis, Dennis. *Making Music: 74 Creative Strategies for Electronic Music Producers.* Berlin: Ableton, 2015.

De Sarkar, Soumavo. "What Is Special or Different about Pro Tools vs Any Other DAW? What Actually Makes It Industry Standard?" *Quora*, June 15, 2020. https://www.quora.com/What-is-special-or-different-about-Pro-Tools-vs-any-other-DAW-What-actually-makes-it-industry-standard.

Diduck, Ryan. *Mad Skills: MIDI and Music Technology in the Twentieth Century.* Marquette, MI: Repeater, 2018.

Dietrich, Frank. "Visual Intelligence: The First Decade of Computer Art (1965–1975)." *Leonardo* 19, no. 2 (1986): 159–169.

"Diggin' in the Carts, Series Trailer." Red Bull Music Academy, August 28, 2014. https://www.youtube.com/watch?v=fwN_o_fi7xE.

Ding, Wei, and Xia Lin. *Information Architecture: The Design and Integration of Information Spaces.* San Rafael, CA: Morgan & Claypool, 2009.

"Diplo about Native Instruments KOMPLETE." Native Instruments, October 19, 2010. https://www.youtube.com/watch?v=9j7Aor_4NTE.

Dolan, Emily. "Toward a Musicology of Interfaces." *Keyboard Perspectives* 5 (2014): 1–14.

Drott, Eric. "Why the Next Song Matters: Streaming, Recommendation, Scarcity." *Twentieth-Century Music* 15, no. 3 (2018): 325–357.

Drummond, Katie. "Army's Virtual Reality Plan: A Digital Doppelganger for Every Soldier." *Wired*, January 18, 2012. http://www.wired.com/dangerroom/2012/01/army-virtual-reality/.

Dyer-Witheford, Nick. "App Worker." In *The Imaginary App*, edited by Paul D. Miller and Svitlana Matviyenko, 125–142. Cambridge, MA: MIT Press, 2014.

Dyson, Frances. *Sounding New Media: Immersion and Embodiment in the Arts and Culture.* Berkeley: University of California Press, 2009.

Dzieza, Josh. "How Hard Will the Robots Make Us Work?" *The Verge*, February 27, 2020. https://www.theverge.com/2020/2/27/21155254/automation-robots-unemployment-jobs-vs-human-google-amazon.

"E3 2011 Machinima Coverage—Sound Shapes on the PS VITA Game Demo & Interview." *Machinima*, June 9, 2011. https://www.youtube.com/watch?v=_rRU0bNeJ20.

Eagleton, Terry. *The Ideology of the Aesthetic*, 2nd ed. Cambridge, MA: Basil Blackwell, 1981.

Eaton, Marcia Muelder. *Basic Issues in Aesthetics.* Belmont, CA: Wadsworth, 1988.

Eidsheim, Nina Sun. *Sensing Sound: Singing and Listening as Vibrational Practice.* Durham, NC: Duke University Press, 2015.

Ekman, Ulrik, ed. *Throughout: Art and Culture Emerging with Ubiquitous Computing.* Cambridge, MA: MIT Press, 2012.

Emmerson, Simon. *Living Electronic Music.* Burlington, VT: Ashgate, 2007.

Eno, Brian. "Generative Music." *In Motion Magazine*, July 7, 1996. http://www.inmotionmagazine.com/eno1.html.

Ensmenger, Nathan. *The Computer Boys Take Over: Computers, Programmers, and the Politics of Technical Expertise.* Cambridge, MA: MIT Press, 2010.

Ensmenger, Nathan. "Making Programming Masculine." In *Gender Codes: Why Women Are Leaving Computing*, edited by Thomas J. Misa, 115–141. Hoboken, NJ: IEEE Computer Society, 2010.

Evans, Claire L. "A Eulogy for Skeuomorphism." *Vice*, June 11, 2013. http://motherboard.vice.com/read/a-eulogy-for-skeumorphism.

Evens, Aden. "Touch in the Abstract." *SubStance* 70, no. 3 (2011): 67–78.

Everett, Anna, and S. Craig Watkins. "The Power of Play: The Portrayal and Performance of Race in Video Games." In *The Ecology of Games: Connecting Youth, Games, and Learning*, edited by Katie Salen, 141–166. Cambridge, MA: MIT Press, 2008.

Ewell, Philip A. "Music Theory and the White Racial Frame." *Music Theory Online* 26, no. 2 (2020). https://mtosmt.org/issues/mto.20.26.2/mto.20.26.2.ewell.html.

Faber, Tom. "Decolonizing Electronic Music Starts with Its Software." *Pitchfork*, February 25, 2021. https://pitchfork.com/thepitch/decolonizing-electronic-music-starts-with-its-software/.

Farman, Jason. *Mobile Interface Theory: Embodied Space and Locative Media*. New York: Routledge, 2012.

Farnell, Andy. "Procedural Audio Theory and Practice." In *The Oxford Handbook of Interactive Audio*, edited by Karen Collins, Bill Kapralos, and Holly Tessler, 531–540. New York: Oxford University Press, 2014.

Farokhmanesh, Megan. "Minecraft Console Sales Surpass PC, 'Almost 54M' Copies Sold in Total." *Polygon*, June 25, 2014. http://www.polygon.com/2014/6/25/5843358/minecraft-console-sales-54m-copies-sold.

Farrugia, Rebekah, and Thomas Swiss. "Tracking the DJs: Vinyl Records, Work, and the Debate over New Technologies." *Journal of Popular Music Studies* 17, no. 1 (2005): 30–44.

Feldman, Jessica. "The MIDI Effect." Paper presented at Bone Flute to Auto-Tune: A Conference on Music & Technology in History, Theory and Practice, Berkeley, CA, April 24–26, 2014.

"Figure for iPhone and iPad." Reason Studios, April 4, 2012. https://www.youtube.com/watch?v=gLLjRH6GJec. "Flow Machines: AI Assisted Music." Sony, n.d. https://www.flow-machines.com.

Finley, Klint. "Obama Becomes First President to Write a Computer Program." *Wired*, December 8, 2014. https://www.wired.com/2014/12/obama-becomes-first-president-write-computer-program/.

Folkestad, G., D. J. Hargreaves, and B. Lindstrom. "A Typology of Composition Styles." *British Journal of Music Education* 15, no. 1 (1998): 83–97.

Foster, Hal. *Design and Crime (and Other Diatribes)*. London: Verso, 2002.

Foucault, Michel. *The Archeology of Knowledge*. Translated by R. Sheridan. New York: Routledge, 1972.

Franinović, Karmen, and Stefania Serafin. *Sonic Interaction Design*. Cambridge, MA: MIT Press, 2013.

Frederick, Matthew. *101 Things I Learned in Architecture School*. Cambridge: MIT Press, 2007.

Frith, Simon, and Simon Zagorski-Thomas, eds. *The Art of Record Production: An Introductory Reader for a New Academic Field*. Burlington, VT: Ashgate, 2012.

Fuller, Matthew. *Media Ecologies: Materialist Energies in Art and Technoculture*. Cambridge, MA: MIT Press, 2007.

Fuller, Matthew, ed. *Software Studies: A Lexicon*. Cambridge, MA: MIT Press, 2008.

Future Music. "What Is Machine Learning, and What Does It Mean for Music?" *Music Radar*, August 30, 2019. https://www.musicradar.com/news/what-is-machine-learning-and-what-does-it-mean-for-music.

Gadir, Tami. "Resistance or Reiteration? Rethinking Gender in DJ Cultures." *Contemporary Music Review* 35, no. 1 (2016): 115–129.

Galloway, Alexander R. *The Interface Effect*. Malden, MA: Polity, 2012.

Galuszka, Patryk. "Netlabels and Democratization of the Recording Industry." *First Monday* 17, no. 7 (July 2012): n.p. http://firstmonday.org/article/view/3770/3278.

Ganszyniec, Artur. "In Praise of Slow Games." *Gamasutra*, June 24, 2019. https://www.gamasutra.com/blogs/ArturGanszyniec/20190624/345350/In_Praise_of_Slow_Games.php.

Gardner, Martin. "Mathematical Games: The Fantastic Combinations of John Conway's New Solitaire Game 'Life.'" *Scientific American* 223, no. 4 (1970): 120–123.

Gaston-Bird, Leslie. *Women in Audio*. London and New York: Routledge, 2020.

Gatlin, Jonathan. *Bill Gates: The Path to the Future*. New York: HarperCollins, 1999.

Geertz, Clifford. "Thick Description: Toward an Interpretative Theory of Culture." In *The Interpretation of Cultures*, 3–36. New York: Basic Books, 1973.

Georgescu, Vlad. "Feature Creep Is a Problem. Learn How to Avoid It." *Plutora*, May 12, 2019. https://www.plutora.com/blog/feature-creep-problem.

Gibson, James J. *The Ecological Approach to Visual Perception*. Hillsdale, NJ: Lawrence Erlbaum, 1986.

Gigliotti, Carol. "Aesthetics of a Virtual World." *Leonardo* 28, no. 4 (1995): 289–295.

Gilmore, Lauren. "The Gender Gap in Technology Can Byte Me." *The Next Web*, December 1, 2015. http://thenextweb.com/insider/2015/12/01/the-gender-gap-in-technology-can-byte-me/#gref.

Gitelman, Lisa. "Media, Materiality, and the Measure of the Digital: Or, the Case of Sheet Music and the Problem of Piano Rolls." In *Memory Bytes: History, Technology, and Digital Culture*, edited by Lauren Rabinovitz and Abraham Geil, 199–217. Durham, NC: Duke University Press, 2004.

Gleick, James. *The Information: A History, A Theory, A Flood*. New York: Pantheon, 2011.

Goehr, Lydia. *The Imaginary Museum of Musical Works: An Essay in the Philosophy of Music*. Oxford: Oxford University Press, 1992.

Golden, Ean. "Music Maneuvers: Discover the Digital Turntablism Concept, 'Controllerism,' Compliments of Moldover." *Remix*, October 2007. http://www.moldover.com/press/Moldover_Remix_Oct-2007_w.jpg.

Goldmann, Stefan. *Presets: Digital Shortcuts to Sound*. Berlin: Macro, 2019.

Goldschmitt, Kariann, and Nick Seaver. "Shaping the Stream: Techniques and Troubles of Algorithmic Recommendation." In *Cambridge Companion to Music and Digital Culture*, edited by Nicholas Cook, Monique Ingalls, David Trippett, and Peter Webb, 298–326. Cambridge: Cambridge University Press, 2019.

Golumbia, David. *The Cultural Logic of Computation*. Cambridge, MA: Harvard University Press, 2009.

Google. "Magenta Studio (v1.0)." Magenta, n.d. https://magenta.tensorflow.org/studio.

Google. "Introduction: Material Design." Google, n.d. https://material.io/design/introduction.

Google. "The Basics of Micro-Moments." Google, n.d. https://www.thinkwithgoogle.com/consumer-insights/consumer-journey/micro-moments-understand-new-consumer-behavior/.

Gracyk, Theodore. *Rhythm and Noise: An Aesthetics of Rock*. Durham, NC: Duke University Press, 1996.

Green, Lucy. "Music Education, Cultural Capital, and Social Group Identity." In *The Cultural Study of Music: A Critical Introduction*, edited by Martin Clayton, Trevor Herbert, and Richard Middleton, 263–273. New York: Routledge, 2003.

Green, Lucy. *Music, Gender, Education*. Cambridge: Cambridge University Press, 1997.

Greher, Gena R., and Jesse M. Heines. *Computational Thinking in Sound: Teaching the Art and Science of Music and Technology*. New York: Oxford University Press, 2014.

Guarnieri, Mya. "Stop Saying 'We're All in This Together.' You Have Money. It's Not the Same." *Washington Post*, April 18, 2020. https://www.washingtonpost.com/outlook/2020/04/18/coronavirus-retail-jobs-inequality/.

Gube, Jacob. "What Is User Experience Design? Overview, Tools and Resources." *Smashing Magazine*, October 5, 2010. http://uxdesign.smashingmagazine.com/2010/10/05/what-is-user-experience-design-overview-tools-and-resources/.

Guevara-Villalobos, Orlando. "Cultures of Independent Game Production: Examining the Relationship between Community and Labour." In *Think Design Play: Proceedings of the 2011 Digital Games Research Association Conference*. Utrecht: DIGRA, 2011.

Hamilton, Robert, Jeffrey Smith, and Ge Wang. "Social Composition: Musical Data Systems for Expressive Mobile Music." *Leonardo Music Journal* 21 (2011): 57–64.

Hands, Joss. "Introduction: Politics, Power, and 'Platformativity.'" *Culture Machine* 14 (2013): 1–9.

Harvey, Steve. "Virtue and Vice and Gear Lust in Brooklyn." *Pro Sound News*, March 2018, n.p.

Haworth, Christopher. "Sound Synthesis Procedures as Texts: An Ontological Politics in Electroacoustic and Computer Music." Computer Music Journal 39, no. 1 (2015): 41–58.

Halberstam, Jack. *The Queer Art of Failure*. Durham, NC: Duke University Press, 2011.

Harkins, Paul. "Microsampling: From Akufen's Microhouse to Todd Edwards and the Sound of UK Garage." In *Musical Rhythm in the Age of Digital Reproduction*, edited by Anne Danielsen, 177–194. Burlington, VT: Ashgate, 2012.

Harper, Richard H. R. *Texture: Human Expression in the Age of Communications Overload*. Cambridge, MA: MIT Press, 2010.

Hayles, N. Katherine. *Electronic Literature: New Horizons for the Literary*. Notre Dame, IN: University of Notre Dame Press, 2008.

Hayles, N. Katherine. *How We Became Posthuman: Virtual Bodies in Cybernetics, Literature, and Informatics*. Chicago: University of Chicago Press, 1999.

Heatherly, Ben, and Logan Howard. "Video Game Controllers." Clemson University. http://andrewd.ces.clemson.edu/courses/cpsc414/spring14/papers/group2.pdf.

Heidegger, Martin. "The Question Concerning Technology." In *Basic Writings*, edited by David Farrell Krell, 3–35. New York: Harper & Row, 1977.

Helmreich, Stefan. "An Anthropologist Underwater: Immersive Soundscapes, Submarine Cyborgs, and Transductive Ethnography." *American Ethnologist* 34 no. 4 (2007): 621–641.

Herber, Norbert. "The Composition-Instrument: Emergence, Improvisation, and Interaction in Games and New Media." In *From Pac-Man to Pop Music: Interactive Audio in Games and New Media*, edited by Karen Collins, 103–123. Burlington, VT: Ashgate, 2008.

Hodges, Richard. "The New Technology." In *Music Education: Trends and Issues*, edited by Charles Plummeridge. London: Institute of Education, University opf London, 1996.

Holmes, Alex. "Review: iZotope Ozone 9." *MusicTech*, October 29, 2019. https://www.musictech.net/reviews/plug-ins/izotope-ozone-9/.

Hookway, Branden. *Interface*. Cambridge: MIT Press, 2014.

Horsfield, Peter. "Continuities and Discontinuities in Ethical Reflections on Digital Virtual Reality." *Journal of Mass Media Ethics* 34 nos. 3-4 (2003): 156-173.

"How Diplo Is Making Pop Music Weird." *Time*, August 20, 2015. https://www.youtube.com/watch?v=lxRDCsYAX50.

"How IBM Watson Inspired Alex Da Kid's New Song 'Not Easy.'" *Business Insider*, October 25, 2016. https://www.businessinsider.com/sc/ibm-watson-helps-create-alex-da-kid-song-2016-10.

Huhtamo, Erkki. *Illusions in Motion: Media Archeology of the Moving Panorama and Related Spectacles.* Cambridge, MA: MIT Press, 2013.

Huizinga, Johan. *Homo Ludens: A Study of the Play-Element in Culture.* Boston: Beacon, 1955.

"The Human Voice Is Human on the New Orthophonic Victrola." David Sarnoff Library digital archive, March 27, 1927. http://digital.hagley.org/cdm/ref/collection/p16038coll11/id/12569.

Hunley, Alex. "Hamlet's BlackBerry—William Powers." *Full Stop*, August 24, 2011. http://www.full-stop.net/2011/08/24/reviews/alexhunley/hamlets-blackberry-william-powers/.

"IBM Watson Beat." IBM, n.d. Accessed August 2, 2020. https://www.ibm.com/watson/music/uk-en/.

Image-Line. "Company History." http://www.image-line.com/company-history/.

Imamović, Maisa. "How to Nothing." Institute of Network Cultures, January 14, 2019. https://networkcultures.org/longform/2019/01/14/how-to-nothing/.

"The Incredible, Amazing, Awesome Apple Keynote." Justanotherguy84, September 15, 2009. https://www.youtube.com/watch?v=Nx7v815bYUw.

Ingold, Tim. *The Perception of the Environment.* New York: Routledge, 2000.

Ingram, Matthew. "Switched-On." *Loops* 1, 2009, 129–137.

"Innovating for Writers and Artists." Spotify, July 12, 2017. https://artists.spotify.com/blog/innovating-for-writers-and-artists.

"Insight 2." iZotope, n.d. https://www.izotope.com/en/products/insight.html.

"iPhone 3G Commercial 'There's an App for That' 2009." CommercialsKid, February 4, 2009. https://www.youtube.com/watch?v=szrsfeyLzyg.

"Is It Live or Is It Memorex." Ukclassictelly, May 14, 2012. https://www.youtube.com/watch?v=lhfugTnXJV4.

"Is It Time for Avid to Refresh the Pro Tools Graphical User Interface?—Poll." *Pro Tools Expert*, May 13, 2019. https://www.pro-tools-expert.com/home-page/2019/4/25/is-it-time-for-avid-to-overhaul-the-entire-pro-tools-user-interface-for-the-future.

Jacko, Julie A. *The Human-Computer Interaction Handbook*, 3rd ed. Boca Raton, FL: CRC, 2012.

Jackson, Gita. "'Animal Crossing: New Horizons' Is Not the Game We All Need Right Now." *Vice*, April 14, 2020. https://www.vice.com/en_us/article/jgezwk/animal-crossing-new-horizons-is-not-the-game-we-all-need-right-now.

Jackson, Reed. "The Story of FruityLoops: How a Belgian Porno Game Company Employee Changed Modern Music." *Vice*, December 11, 2015. https://noisey.vice.com/en_us/article/rnwkvz/fruity-loops-fl-studio-program-used-to-create-trap-music-sound.

Jacob, Robert J. K., Audrey Girouard, Leanne M. Hirshfield, Michael S. Horn, Orit Shaer, Erin Treacy Solovey, and Jamie Zigelbaum. "Reality-Based Interaction: A Framework for Post-WIMP Interfaces." In *CHI 2008 Proceedings: Post-WIMP*, 201–210. Florence: ACM, 2008.

James, Robin. "Aesthetics and Live Electronic Music Performance." *Cyborgology*, December 13, 2013. http://thesocietypages.org/cyborgology/2013/12/13/aesthetics-and-live-electronic-music-performance/.

James, Robin. "Poptimism and Popular Feminism." *Sounding Out!*, September 17, 2018. https://soundstudiesblog.com/2018/09/17/poptimism-and-popular-feminism/.

James, Robin. *Resilience & Melancholy: Pop Music, Feminism, Neoliberalism.* Alresford: Zero Books, 2015.

James, Robin. *The Sonic Episteme: Acoustic Resonance, Neoliberalism, and Biopolitics.* Durham, NC: Duke University Press, 2019.

Janlert, Lars-Erik, and Erik Stolterman. *The Things That Keep Us Busy: The Elements of Interaction.* Cambridge, MA: The MIT Press, 2017.

Jenkins, Henry. *Confronting the Challenges of Participatory Culture: Media Education for the 21st Century.* Cambridge, MA: MIT Press, 2009.

Jennemann, Annie. "'Animal Crossing: New Horizons' Is the Perfect Game for Self-Isolating." *Vox*, March 26, 2020. https://www.voxmagazine.com/arts/animal-crossing-new-horizons-is-the-perfect-game-for-self-isolating/article_84d78f9a-6ee7-11ea-a7c0-4f0900a6b9a6.html.

Jesse! "FL Studio Is . . . Compared to Pro Tools?" Image-Line forum, February 28, 2012. https://forum.image-line.com/viewtopic.php?f=100&t=89594.

Johnson, Steven. *Interface Culture.* New York: HarperCollins, 1997.

Juul, Jesper. *A Casual Revolution: Reinventing Video Games and Their Players.* Cambridge, MA: MIT Press, 2010.

Kachka, Boris, Rebecca Milzoff, and Dan Reilly. "10 Tricks That Musicians and Actors Use in Live Performances." *Vulture*, April 3, 2016. http://www.vulture.com/2016/03/live-performance-tricks.html.

Kane, Brian. *Sound Unseen: Acousmatic Sound in Theory and Practice.* New York: Oxford University Press, 2014.

Kant, Immanuel. *Critique of Judgment.* Translated by Werner S. Pluhar. Indianapolis: Hackett, 1987.

Kartomi, Margaret. "The Classification of Musical Instruments: Changing Trends in Research from the Late Nineteenth Century, with Special Reference to the 1990s." *Ethnomusicology* 45, no. 2 (2001): 283–314.

Kassabian, Anahid. *Ubiquitous Listening: Affect, Attention, and Distributed Subjectivity.* Berkeley: University of California Press, 2013.

Katz, Mark. "The Amateur in the Age of Mechanical Music." In *The Oxford Handbook of Sound Studies*, edited by Trevor Pinch and Karin Bijsterveld, 459–479. New York: Oxford University Press, 2012.

Katz, Mark. *Capturing Sound: How Technology Has Changed Music.* Berkeley: University of California Press, 2004.

Katz, Mark. *Groove Music: The Art and Culture of the Hip-Hop DJ.* New York: Oxford University Press, 2012.

Kay, Alan. "Computer Software." *Scientific American* 251, no. 3 (1984): 52–59.

Keil, Charles. "The Theory of Participatory Discrepancies: A Progress Report." *Ethnomusicology* 39, no. 1 (1995): 1–19.

Kelly, Heather. "Do Video Games Cause Violence?" CNN, August 17, 2015. http://money.cnn.com/2015/08/17/technology/video-game-violence/.

Kelty, Christopher M. *Two Bits: The Cultural Significance of Free Software.* Durham, NC: Duke University Press, 2008.

Khan, Imad. "Why Animal Crossing Is the Game for the Coronavirus Moment." *New York Times*, April 7, 2020. https://www.nytimes.com/2020/04/07/arts/animal-crossing-covid-coronavirus-popularity-millennials.html.

Kiesler, S., L. Sproull, and J. S. Eccles. "Pool Halls, Chips and War Games: Women in the Culture of Computing." *Psychology of Women Quarterly* 9 (1985): 451–462.

Kincaid, Jason. "RjDj Generates an Awesome, Trippy Soundtrack for Your Life." *TechCrunch*, October 13, 2008. http://techcrunch.com/2008/10/13/rjdj-generates-an-awesome-trippy-soundtrack-for-your-life/.

Kirn, Peter. "How to Get into a Creative Flow with FL Studio—and What Could Make It Worth It." *CreateDigitalMusic*, June 10, 2020. https://cdm.link/2020/06/how-to-get-into-a-creative-flow-with-fl-studio/.

Kirn, Peter. "Live Music Tools at the Bleeding Edge: Laura Escudé, from Herbie Hancock Collabs to Solo (Live Berlin, Sat)." *CDM*, July 27, 2012. https://cdm.link/2012/07/live-music-tools-at-the-bleeding-edge-laura-escude-from-herbie-hancock-collabs-to-solo-live-berlin-sat/.

Kisner, Jordan. "The Dark Art of Mastering Music." *Pitchfork*, May 19, 2016. https://pitchfork.com/features/article/9894-the-dark-art-of-mastering-music/.

Kittler, Friedrich. *Discourse Networks 1800/1900*. Stanford, CA: Stanford University Press, 1990.

Kittler, Friedrich. *Gramophone, Film, Typewriter*. Translated by Geoffrey Winthrop-Young and Michael Wutz. Stanford, CA: Stanford University Press, 1999.

Kittler, Friedrich. "There Is No Software." In *The Truth of the Technological World: Essays on the Genealogy of Presence*, 219–229. Stanford, CA: Stanford University Press, 2014.

Klein, Naomi. "Coronavirus Capitalism—and How to Beat It." *The Intercept*, March 16, 2020. https://theintercept.com/2020/03/16/coronavirus-capitalism/.

Koloskus, Jack. "Apple, Big Sur, and the Rise of Neumorphism." *Input*, June 24, 2020. https://www.inputmag.com/design/apple-macos-big-sur-the-rise-of-neumorphism.

Koutsomichalis, Marinos. "From Music to Big Music: Listening in the Age of Big Data." *Leonardo Music Journal* 26 (2016): 24–27.

Krug, Steve. *Don't Make Me Think, Revisited: A Common Sense Approach to Web and Mobile Usability*. San Francisco: New Riders, 2014.

Laine, Rachel. "Syn & Synthesizers." Flickr. https://www.flickr.com/photos/bifrostgirl/albums/72157649002021637.

Langton, Christopher. "Artificial Life." In *Artificial Life*, edited by Christopher Langton, 1–47. New York: Addison-Wesley, 1989.

Larson, Brad. "Rapid Prototyping in Swift Playgrounds." *objc.io* 16 (September 2014): n.p. https://www.objc.io/issues/16-swift/rapid-prototyping-in-swift-playgrounds/.

Larson, Magali Sarfatti. *The Rise of Professionalism: Monopolies of Competence and Sheltered Markets*. Livingston, NJ: Transaction, 2012.

"Las Vegas." *Noisey*. Vice TV, April 7, 2016.

Latour, Bruno. *Reassembling the Social: An Introduction to Actor-Network Theory*. Oxford: Oxford University Press, 2005.

Latour, Bruno. "Technology Is Society Made Durable." In *A Sociology of Monsters? Essays on Power, Technology and Domination*, edited by John Law, 103–131. London: Routledge, 1991.

"Laura Escudé: The Art of Designing a Live Show." Ableton, September 24, 2019. https://www.ableton.com/en/blog/laura-escude-art-designing-live-show/.

Laurel, Brenda. *Computers as Theatre*. Boston: Addison-Wesley, 1993.

Ledesma, Eduardo. "The Poetics and Politics of Computer Code in Latin America: Codework, Code Art, and Live Coding." *Revista de Estudios Hispánicos* 49, no. 1 (2015): 91–120.

Legg, Robert. "'One Equal Music': An Exploration of Gender Perceptions and the Fair Assessment by Beginning Music Teachers of Musical Compositions." *Music Education Research* 12, no. 2 (2010): 141–149.

Le Guin, Elisabeth. *Boccherini's Body: An Essay in Carnal Musicology*. Berkeley: University of California Press, 2006.

Le Guin, Elisabeth. "'Cello-and-Bow Thinking': Boccherini's Cello Sonata in Eb Major, 'Fuori Catalog.'" *Echo* 1, no. 1 (1999). http://www.echo.ucla.edu/Volume1-Issue1/leguin/leguin-article.html.

Leman, Marc. *Embodied Music Cognition and Mediation Technology*. Cambridge, MA: MIT Press, 2008.

Leppert, Richard. "Music 'Pushed to the Edge of Existence' (Adorno, Listening, and the Question of Hope)." *Cultural Critique* 60 (2005): 92–133.

Lessig, Lawrence. *Remix: Making Art and Commerce Thrive in the Hybrid Economy*. New York: Penguin, 2008.

Levitz, Tamara. "The Musicological Elite." *Current Musicology* 102 (2018): 9–49.

Levy, Steven. *Hackers: Heroes of the Computer Revolution*. Garden City, NY: Doubleday, 1984.

Licklider, J. C. R. "Man-Computer Symbiosis." *IRE Transactions on Human Factors in Electronics* 1 (1960): 4–11.

Light, Ann. "Performing Interaction Design with Judith Butler." In *Critical Theory and Interaction Design*, edited by Jeffrey Bardzell, Shaowen Bardzell, and Mark Blythe, 429–446. Cambridge, MA: MIT Press, 2018.

Lin, Rong-Gong, II. "ER Doctors: Drug-Fueled Raves Too Dangerous and Should Be Banned." *Los Angeles Times*, August 10, 2015. http://www.latimes.com/local/lanow/la-me-ln-why-some-er-doctors-want-to-end-raves-in-los-angeles-county-20150810-story.html.

Lipkin, Nadav. "Examining Indie's Independence: The Meaning of 'Indie' Games, the Politics of Production, and Mainstream Co-optation." *Loading* 7, no. 11 (2012): 8–24.

"Livecoding.tv: Watch People Code Products Live." Education Ecosystem Blog, February 9, 2015. https://blog.education-ecosystem.com/livecoding-tv-watch-people-code-products-live/.

Lovink, Geert. *Dynamics of Critical Internet Culture (1994–2001)*. Amsterdam: Institute of Network Cultures, 2009.

Lunau, Kate. "Google's Ray Kurzweil on the quest to live forever." *Maclean's*, October 14, 2013. https://www.macleans.ca/society/life/how-nanobots-will-help-the-immune-system-and-why-well-be-much-smarter-thanks-to-machines-2/.

Lunenfeld, Peter. "The Maker's Discourse." Critical Mass: The Legacy of Hollis Frampton, Invited Conference. University of Chicago, 2010.

Lunenfeld, Peter. "Mia Laboro: Maker's Envy and the Generative Humanities." Book Presence in a Digital Age Conference. University of Utrecht, 2012.

Lunenfeld, Peter. *The Secret War between Downloading and Uploading: Tales of the Computer as Culture Machine*. Cambridge, MA: MIT Press, 2011.

Lunning, Frenchy, ed. *Mechademia 5: Fanthropologies*. Minneapolis: University of Minnesota Press, 2010.

Luta, Primus. "Musical Objects, Variability and Live Electronic Performance." *Sounding Out!*, August 12, 2013. http://soundstudiesblog.com/2013/08/12/musical-objects-variability/.

Luta, Primus. "Toward a Practical Language for Live Electronic Performance." *Sounding Out!*, April 29, 2013. http://soundstudiesblog.com/2013/04/29/toward-a-practical-language-for-live-electronic-performance/.

Lyon, Eric. *Designing Audio Objects for Max/MSP and Pd*. Middleton, WI: A-R Editions, 2012.

MacColl, Ian, and Ingrid Richardson. "A Cultural Somatics of Mobile Media and Urban Screens: Wiffiti and the IWALL Prototype." *Journal of Urban Technology* 15, no. 3 (2008): 99–116.

Machkovech, Sam. "SoundCloud's Free 'Auto-Mastering' Audio Tool Is More of an Auto-Turd." *Ars Technica*, May 29, 2016. https://arstechnica.com/gadgets/2016/05/soundclouds-free-auto-mastering-audio-tool-is-more-of-an-auto-turd/.

Mackenzie, Adrian. "Transduction: Invention, Innovation and Collective Life." March 20, 2003. http://www.lancs.ac.uk/staff/mackenza/papers/transduction.pdf.

Macmillan, Fiona, ed. *New Directions in Copyright Law*, Vol. 1. Northampton, MA: Edward Elgar, 2005.Magnusson, Thor. "Affordances and Constraints in Screen-Based Musical Instruments." In *Proceedings of the 4th Nordic Conference on Human-Computer Interaction: Changing Roles*, 441–444. New York: ACM, 2006.

Magnusson, Thor. "Algorithms as Scores: Coding Live Music." *Leonardo Music Journal* 21 (2011): 19–23.

Magnusson, Thor. *Sonic Writing: Technologies of Material, Symbolic, and Signal Inscriptions*. New York: Bloomsbury, 2019.

"Making the iPhone: The Design Tao of Apple's Jony Ive." Bloomberg, September 19, 2015. https://www.youtube.com/watch?v=e9OJMlVQBiY.

Mandelbaum, Melissa. "Applying Architecture to Product Design: Parti." *Medium*, March 21, 2016. https://medium.com/applying-architecture-to-product-design/applying-architecture-to-product-design-parti-68b0b7345db9#.p0d87v6lc.

Manovich, Lev. *The Language of New Media*. Cambridge, MA: MIT Press, 2001.

Manovich, Lev. *Software Takes Command*. New York: Bloomsbury, 2013.

Martin, Chase Bowen, and Mark Deuze. "The Independent Production of Culture: A Digital Games Case Study." *Games and Culture* 4, no. 3 (2009): 276–295.

Massumi, Brian. *Parables for the Virtual: Movement, Affect, Sensation*. Durham, NC: Duke University Press, 2002.

Mateas, Michael. "Procedural Literacy: Educating the New Media Practitioner." In *Beyond Fun: Serious Games and Media*, edited by Drew Davidson, 67–83. Pittsburgh: ETC, 2008. http://citeseerx.ist.psu.edu/viewdoc/download?doi=10.1.1.90.2106&rep=rep1&type=pdf.

Matviyenko, Svitlana. "Introduction." In *The Imaginary App*, edited by Paul D. Miller and Svitlana Matviyenko, xvii–xxxvi. Cambridge, MA: MIT Press, 2014.

Mauro-Flude, Nancy. "The Intimacy of the Command Line." *Scan* 10, no. 2 (2013): n.p. http://scan.net.au/scn/journal/vol10number2/Nancy-Mauro-Flude.html.

May, Tom. "The Beginner's Guide to Flat Design." *Creative Bloq*, January 14, 2021. https://www.creativebloq.com/graphic-design/what-flat-design-3132112.

McAllister, Peter. *Manthropology: The Science of Why the Modern Male Is Not the Man He Used to Be*. New York: St. Martin's, 2010.

McCarthy, Erin. *Ethics Embodied: Rethinking Selfhood through Continental, Japanese, and Feminist Philosophies*. Lanham, MD: Rowman & Littlefield, 2010.

McGann, Jerome. *Radiant Textuality: Literature after the World Wide Web*. New York: Palgrave, 2001.

McGonigal, Jane. *Reality Is Broken: Why Games Make Us Better and How They Can Change the World*. New York: Penguin, 2011.

McKittrick, Katherine, and Alexander G. Weheliye. "808s & Heartbreak." *Propter Nos* 2, no. 1 (2017): 13–42.

McLean, Alex, Julian Rohrhuber, and Nick Collins, eds. Special Issue on Live Coding. *Computer Music Journal* 38, no. 1 (2014).

McLuhan, Marshall. *The Medium Is the Massage: An Inventory of Effects*. Corte Madera, CA: Gingko, 2001.

McLuhan, Marshall. *Understanding Media: The Extensions of Man*. New York: McGraw-Hill, 1964.

McNally, James. "Favela Chic: Diplo, *Funk Carioca*, and the Ethics and Aesthetics of the Global Remix." *Popular Music and Society* 40, no. 4 (2016): 434–452.

McPherson, Tara. "Why Are the Digital Humanities So White? Or Thinking the Histories of Race and Computation." In *Debates in the Digital Humanities*, edited by Matthew K. Gold, 139–160. Minneapolis: University of Minnesota Press, 2012.

Mdb. "You Don't Need Pro Tools Dude!" Ableton Forum, June 18, 2006. https://forum.ableton.com/viewtopic.php?t=40683.

Mechaber, Ezra. "President Obama Is the First President to Write a Line of Code." White House Blog, December 10, 2014. http://www.whitehouse.gov/blog/2014/12/10/presid ent-obama-first-president-write-line-code.

"Meet Liquid Rhythm." WaveDNA, n.d. https://www.wavedna.com/liquid-rhythm/.

Merrin, William. "The Rise of the Gadget and Hyperludic Me-dia." *Cultural Politics* 10, no. 1 (2014): 1–20.

Microsoft. "Get Your Start with an Hour of Code." N.d. http://www.microsoft.com/about/corporatecitizenship/en-us/youthspark/youthsparkhub/hourofcode/.

Milam, Erika Lorraine, and Robert A. Nye. "An Introduction to *Scientific Masculinities*." *Osiris* 30, no. 1 (2015): 1–14.

Millard, Drew. "Interview: Rustie Used to Produce Like How Gamers Game." *Pitchfork*, May 31, 2012. http://archive.is/dzPoJ.

Miller, Kiri. *Playable Bodies: Dance Games and Intimate Media*. New York: Oxford University Press, 2017.

Miller, Kiri. *Playing Along: Digital Games, YouTube, and Virtual Performance*. New York: Oxford University Press, 2012.

Mills, Mara. "Do Signals Have Politics? Inscribing Abilities in Cochlear Implants." In *The Oxford Handbook of Sound Studies*, edited by Trevor Pinch and Karin Bijsterveld, 320–346. Oxford: Oxford University Press, 2012.

Mishra, Pankaj. "The Crisis in Modern Masculinity." *Guardian*, March 17, 2018. https://www.theguardian.com/books/2018/mar/17/the-crisis-in-modern-masculinity.

"Moldover: Performance and Controllerism." Ableton, November 28, 2013. https://www.ableton.com/en/blog/moldover-performance-and-controllerism/.

Molina, Brett. "Obama Seeks Research into Violent Video Games." *USA Today*, January 16, 2013. http://www.usatoday.com/story/tech/gaming/2013/01/16/obama-gun-viole nce-video-games/1839879/.

Montano, Ed. "'How Do You Know He's Not Playing Pac-Man While He's Supposed to Be DJing?': Technology, Formats and the Digital Future of DJ Culture." *Popular Music* 29, no. 3 (2010): 397–416.

Montfort, Nick, and Ian Bogost. *Racing the Beam: The Atari Video Computer System.* Cambridge, MA: MIT Press, 2009.

Montfort, Nick, Patsy Baudoin, John Bell, Ian Bogost, Jeremy Douglass, Mark C. Marino, Michael Mateas, Casey Reas, Mark Sample, and Noah Vawter. *10 PRINT CHR$(205.5+ RND(1)); : GOTO 10.* Cambridge, MA: MIT Press, 2013.

"Monu/Mental." Unity3D, January 23, 2014. https://unity3d.com/showcase/case-stories/monument-valley.

Morozov, Evgeny. *To Save Everything, Click Here: The Folly of Technological Solutionism.* New York: Public Affairs, 2013.

Morrogh, Earl. *Information Architecture: An Emerging 21st Century Profession.* Upper Saddle River, NJ: Prentice Hall, 2003.

Moseley, Roger. "Playing Games with Music (and Vice Versa): Ludomusicological Perspectives on *Guitar Hero* and *Rock Band.*" In *Taking It to the Bridge: Music as Performance,* edited by Nicholas Cook and Richard Pettengill, 279–318. Ann Arbor: University of Michigan Press, 2013.

Moulier-Boutang, Yann. *Cognitive Capitalism.* Boston: Polity, 2011.

Muggs, Joe. "Flying Lotus: Cosmic Drama." *Resident Advisor,* March 24, 2010. http://www.residentadvisor.net/feature.aspx?1175.

Murray, Janet. *Inventing the Medium: Principles of Interaction Design as a Cultural Practice.* Cambridge, MA: MIT Press, 2012.

Myers, David. "The Video Game Aesthetic: Play as Form." In *Video Game Theory Reader 2,* edited by Bernard Perron and Mark J. P. Wolf. New York: Routledge, 2009.

Nadin, Mihai. "Emergent Aesthetics: Aesthetic Issues in Computer Arts." *Leonardo* 2 (1989): 43–48.

Naji, Cassandra, and Emily Grace Adiseshiah. *How Top Companies Prototype Killer Products.* San Francisco, CA: Justinmind, 2017.

Nash, Chris, and Alan F. Blackwell. "Flow of Creative Interaction with Digital Music Notations." In *The Oxford Handbook of Interactive Audio,* edited by Karen Collins, Bill Kapralos, and Holly Tessler, 387–404. New York: Oxford University Press, 2013.

Native Instruments. "Discovery Series: Collection." https://www.native-instruments.com/en/products/komplete/world/discovery-series-collection/.

Native Instruments. "Maschine Studio." February 1, 2016. http://www.native-instruments.com/en/products/maschine/production-systems/maschine-studio/.

Nercessian, Shahan. "iZotope and Assistive Audio Technology." iZotope, July 19, 2018. https://www.izotope.com/en/learn/izotope-and-assistive-audio-technology.html.

Nevile, Ben. "An Interview with Kim Cascone." Cycling '74, September 13, 2005. https://cycling74.com/2005/09/13/an-interview-with-kim-cascone/.

Ng, Valerie, and Jon Tilliss. "Balancing Aesthetics and Usability in Medical User Interface Design: 9 Key Trends." *Emergo,* July 8, 2018. https://www.emergobyul.com/blog/2018/07/balancing-aesthetics-and-usability-medical-user-interface-design-9-key-trends.

Ngak, Chenda. "Tupac Coachella Hologram: Behind the Technology." CBS News, November 9, 2012. http://www.cbsnews.com/news/tupac-coachella-hologram-behind-the-technology/.

"1984 Apple's Macintosh Commercial (HD)." Mac History, February 1, 2012. https://www.youtube.com/watch?v=VtvjbmoDx-I.

Norman, Donald A. *The Design of Everyday Things*. New York: Basic Books, 1988.

Nusselder, André. *Interface Fantasy: A Lacanian Cyborg Ontology*. Cambridge, MA: MIT Press, 2009.

Nystrom, Christine. "Towards a Science of Media Ecology: The Formulation of Integrated Conceptual Paradigms for the Study of Human Communication Systems." PhD diss., New York University, 1973.

Objectified. Directed by Gary Hustwit. Plexi Productions, 2009.

Oldenzeil, Ruth. *Making Technology Masculine: Men, Women, and Modern Machines in America, 1870–1945*. Amsterdam: Amsterdam University Press, 2004.

Oliveros, Pauline. *Deep Listening: A Composer's Sound Practice*. Lincoln, NE: iUniverse, 2005.

O'Neill, Cathy. *Weapons of Mass Destruction: How Big Data Increases Inequality and Threatens Democracy*. New York: Crown, 2016.

O'Neill, Susan A., and Michael J. Boulton. "Boys' and Girls' Preferences for Musical Instruments: A Function of Gender?" *Psychology of Music* 24, no. 2 (1996): 171–183.

"Online Audio Mastering." LANDR, n.d. https://www.landr.com/en/online-audio-mastering.

Orthentix. "The Afflicting Intersections of Gender and Music Production." *Medium*, January 13, 2009. https://medium.com/orthentix/the-afflicting-intersections-of-gender-and-music-production-a0917a41944c.

Ortner, Sherry. *Not Hollywood: Independent Film at the Twilight of the American Dream*. Durham, NC: Duke University Press, 2013.

Owsinski, Bobby. "Is Mastering an Endangered Audio Job?" Music Production Blog, July 10, 2019. https://bobbyowsinskiblog.com/2019/07/10/mastering-endangered/.

Owsinski, Bobby. *The Mastering Engineer's Handbook*, 4th ed. Burbank, CA: Bobby Owsinski Media Group, 2017.

Parisi, David. *Archaeologies of Touch: Interfacing with Haptics from Electricity to Computing*. Minneapolis: University of Minnesota Press, 2018.

Park, Tae Hong. "An Interview with Max Mathews." *Computer Music Journal* 33, no. 3 (2009): 9–22.

Pask, Andrew. "Mini Interview: Jonny Greenwood." Cycling '74, January 2, 2014. https://cycling74.com/2014/01/02/mini-interview-jonny-greenwood/.

Patel, Nilay. "The Technics SL-1200 Turntable Returns in Two New Audiophile Models." *The Verge*, January 5, 2016. http://www.theverge.com/2016/1/5/10718234/technics-sl1200-sl1200g-sl1200gae-turntable-new-models-announced-release-date-ces-2016.

Pattison, Louis. "Video Game Loops & Aliens: Flying Lotus Interviewed." *The Quietus*, May 18, 2010. http://thequietus.com/articles/04269-flying-lotus-interview-cosmogramma.

Payne, Christine McLeavey. "MuseNet." OpenAI, April 25, 2019. https://openai.com/blog/musenet/.

Pearson, Matt. *Generative Art: A Practical Guide Using Processing*. Shelter Island, NY: Manning, 2011.

"Persuasion and the Other Thing: A Critique of Big Data Methodologies in Politics." *Ethnography Matters*, May 24, 2017. https://ethnographymatters.net/blog/2017/05/24/persuasion-and-the-other-thing-a-critique-of-big-data-methodologies-in-politics/.

Peters, Michael. "The Birth of Loop." *Prepared Guitar*, April 25, 2015. http://preparedguitar.blogspot.com/2015/04/the-birth-of-loop-by-michael-peters.html.

Phelan, Peggy. *Unmarked: The Politics of Performance*. New York: Routledge, 1993.

Pinch, Trevor and Karin Bijsterveld, eds. *The Oxford Handbook of Sound Studies*. New York: Oxford University Press, 2012.

Pink, Sarah. *Doing Sensory Ethnography*. London: Sage, 2009.

Polanyi, Michael. *The Tacit Dimension*. Chicago: University of Chicago Press, 1966.

Pold, Søren Bro, and Christian Ulrik Andersen. "Controlled Consumption Culture: When Digital Culture Becomes Software Business." In *The Imaginary App*, edited by Paul D. Miller and Svitlana Matviyenko, 17–33. Cambridge, MA: MIT Press, 2014.

Postman, Neil. "The Reformed English Curriculum." In *High School 1980: The Shape of the Future in American Secondary Education*, edited by Alvin C. Eurich, 160–168. New York: Pitman, 1970.

Powell, Michael. "Obscure Musicology Journal Sparks Battles over Race and Free Speech." *New York Times*, February 14, 2021. https://www.nytimes.com/2021/02/14/arts/mus icology-journal-race-free-speech.html.

Powers, William. *Hamlet's BlackBerry: Building a Good Life in the Digital Age*. New York: Harper Perennial, 2011.

Provenzano, Catherine. "Making Voices: The Gendering of Pitch Correction and the Auto-Tune Effect in Contemporary Pop Music." *Journal of Popular Music Studies* 31, no. 2 (2019): 63–84.

Puckette, Miller. "The Deadly Embrace between Music Software and Its Users." In *Electroacoustic Music beyond Performance: Proceedings of the Electroacoustic Music Studies Network Conference*. Berlin: EMS Network, 2014.

Puckette, Miller. "Max at Seventeen." *Computer Music Journal* 26, no. 4 (2002): 31–43.

Puckette, Miller. "The Patcher." In *Proceedings of the International Computer Music Asssociation*, 420–429. San Francisco: ICMA, 1988. http://msp.ucsd.edu/Publications/ icmc88.pdf.

Radzikowska, Milena, Jennifer Roberts-Smith, Xinyue Zhou, and Stan Ruecker. "A Speculative Feminist Approach to Project Management." *Strategic Design Research Journal* 12, no. 1 (2019): 94–113.

Raessens, Joost. "Computer Games as Participatory Media Culture." In *Handbook of Computer Game Studies*, edited by Joost Raessens and Jeffrey Goldstein, 373–388. Cambridge, MA: MIT Press, 2006.

Raessens, Joost. *Homo Ludens 2.0: The Ludic Turn in Media Theory*. Utrecht: Utrecht University, 2012.

Raessens, Joost. "Playful Identitiers, or the Ludification of Culture." In *Handbook of Computer Game Studies*, edited by Joost Raessens and Jeffrey Goldstein, 373–378. Cambridge, MA: MIT Press, 2006.

Raessens, Joost, and Jeffrey Goldstein. *Handbook of Computer Game Studies*. Cambridge, MA: MIT Press, 2006.

Raskin, Jef. *The Humane Interface: New Directions for Designing Interactive Systems*. Boston: Addison-Wesley Professional, 2000.

Raymond, Eric S. *The Art of Unix Programming*. Boston: Addison-Wesley, 2003.

Reas, Casey, and Chandler McWilliams. *Form+Code in Design, Art, and Architecture*. New York: Princeton Architectural Press, 2010.

Rehn, Alf. "The Politics of Contraband: The Honor Economies of the Warez Scene." *Journal of Socio-Economics* 33 (2004): 359–374.

Reich, Steve. *Writings on Music: 1965–2000*. New York: Oxford University Press, 2002.

"Report: 90% of Waking Hours Spent Staring at Glowing Rectangles." *The Onion*, June 15, 2009. http://www.theonion.com/article/report-90-of-waking-hours-spent-staring-at-glowing-2747.

Resmini, Andrea, and Luca Rosati. *Pervasive Information Architecture: Designing Cross-Channel User Experiences*. Burlington, MA: Morgan Kauffman, 2011.

Reyes, Ian. "To Know beyond Listening: Monitoring Digital Music." *The Senses and Society* 5, no. 3 (2010): 322–338.

Reynaldo, Shawn. "In the Studio: Machinedrum." *XLR8R*, August 27, 2012. https://www.xlr8r.com/gear/2012/08/in-the-studio-machinedrum/.

Reynolds, Simon. "Maximal Nation: Electronic Music's Evolution toward the Thrilling Excess of Digital Maximalism." *Pitchfork*, December 6, 2011. http://pitchfork.com/features/articles/8721-maximal-nation/.

Rhys-Taylor, Alex. "The Essences of Multiculture: A Sensory Exploration of an Inner-City Street Market." *Identities* 20, no. 4 (2013): 393–406.

"Rhythmiq." Accusonus, n.d. https://accusonus.com/products/rhythmiq.

Ribeiro, Márcio, Paulo Borba, and Claus Brabrand. *Emergent Interfaces for Feature Modularization*. Cham, Switzerland: Springer International, 2014.

Richard, Gabriela T., and Kishonna L. Gray. "Gendered Play, Racialized Reality: Black Cyberfeminism, Inclusive Communities of Practice and the Intersections of Learning, Socialization and Resilience in Online Gaming." *Frontiers* 39, no. 1 (2018): 112–148.

Richardson, Ingrid. "Faces, Interfaces, Screens: Relational Ontologies of Framing, Attention and Distraction." *Transformations* 18 (2010). http://www.transformationsjournal.org/wp-content/uploads/2017/01/Richardson_Trans18.pdf.

Richardson, Ingrid. "Ludic Mobilities: The Corporealities of Mobile Gaming." *Mobilities* 5 (2010): 431–447.

Richardson, Ingrid. "Mobile Technosoma: Some Phenomenological Reflections on Itinerant Media Devices." *Fibreculture Journal* 6 (2005). https://six.fibreculturejournal.org/fcj-032-mobile-technosoma-some-phenomenological-reflections-on-itinerant-media-devices/.

Richardson, Ingrid. "Pocket Technospaces: The Bodily Incorporation of Mobile Media." *Continuum* 21 (2007): 205–215.

Richardson, Ingrid. "Touching the Screen: A Phenomenology of Mobile Gaming and the iPhone." In *Studying Mobile Media: Cultural Technologies, Mobile Communication, and the iPhone*, edited by Larissa Hjorth, Jean Burgess, and Ingrid Richardson, 133–153. New York: Routledge, 2012.

Richardson, Ingrid, and Rowan Wilken. "Haptic Vision, Footwork, Place-Making: A Peripatetic Phenomenology of the Mobile Phone Pedestrian." *Second Nature* 1 (2009): 22–41.

Roberts, Eric. "A History of Capacity Challenges in Computer Science." Stanford Computer Science, March 7, 2016. https://cs.stanford.edu/people/eroberts/CSCapacity.pdf.

Rodgers, Tara. "Synthesizing Sound: Metaphor in Audio-Technical Discourse and Synthesis History." PhD diss., McGill University, 2011.

Rodgers, Tara. "Towards a Feminist Historiography of Electronic Music." In *The Sound Studies Reader*, edited by Jonathan Sterne, 475–489. New York: Routledge, 2012.

Roell, Craig H. *The Piano in America, 1890–1940*. Chapel Hill: University of North Carolina Press, 1989.

Romanyshyn, Robert. *Technology as Symptom and Dream*. New York: Routledge, 1989.

Rose, Tricia. *Black Noise: Rap Music and Black Culture in Contemporary America*. Hanover, NH: Wesleyan University Press, 1994.

Rosenberger, Robert. "The Spatial Experience of Telephone Use." *Environment, Space, Place* 2, no. 2 (2010): 61–75.

Rosenfeld, Louis, and Peter Morville. *Information Architecture for the World Wide Web.* Sebastopol, CA: O'Reilly, 1998.

Rovan, Joseph Butch. "Living on the Edge: Alternate Controllers and the Obstinate Interface." In *Mapping Landscapes for Performance as Research: Scholarly Acts and Creative Cartographies*, edited by Shannon Rose Riley and Lynette Hunter, 252–259. New York: Palgrave Macmillan, 2009.

Ruberg, Bonnie. *Video Games Have Always Been Queer.* New York: New York University Press, 2019.

Ruffino, Paolo. "Independent Games: Narratives of Emancipation in Video Game Culture." *Loading* 7, no. 11 (2012): 106–121.

"Rural Legend." Unity3D, April 25, 2014. https://unity3d.com/showcase/case-stories/endless-legend.

"Sakura Tsuruta on Push." Ableton, October 21, 2019. https://www.ableton.com/en/blog/sakura-tsuruta-push/.

Salen, Katie, and Eric Zimmerman. *Rules of Play: Game Design Fundamentals.* Cambridge, MA: MIT Press, 2004.

Salkind, Micah. *Do You Remember House? Chicago's Queer of Color Undergrounds.* New York: Oxford University Press, 2019.

Sallis, Friedemann, ed. *Music Sketches.* Cambridge: Cambridge University Press, 2015.

Sam. "Image Line—Jean-Marie Cannie [CTO and Co-Founder]." *Speakhertz*, May 16, 2013. http://speakhertz.com/16/interview-image-line-co-founder-jean-marie-cannie.

Schloss, Joseph G. *Making Beats: The Art of Sample-Based Hip-Hop.* Middletown, CT: Wesleyan University Press, 2004.

Senior, Mike. *Mixing Secrets for the Small Studio.* New York: Routledge, 2012.

Sharples, M., P. McAndrew, M. Weller, R. Ferguson, E. FitzGerald, T. Hirst, and M. Gaved. "Maker Culture: Learning by Making." In *Innovating Pedagogy 2013: Open University Innovation Report 2*, 33–35. Milton Keynes, UK: Open University, 2013.

Shneiderman, Ben. "Direct Manipulation: A Step beyond Programming Languages." *Computer* 16, no. 8 (1983): 57–69.

Simon, Bart. "Wii Are Out of Control: Bodies, Game Screens and the Production of Gestural Excess." March 5, 2009. https://papers.ssrn.com/sol3/papers.cfm?abstract_id=1354043.

Sinnreich, Aram. *Mashed Up: Music, Technology, and the Rise of Configurable Culture.* Amherst: University of Massachusetts Press, 2010.

Sisario, Ben, and James C. McKinley Jr. "Drug Deaths Threaten Rising Business of Electronic Music Fests." *New York Times*, September 9, 2013. http://www.nytimes.com/2013/09/10/arts/music/drugs-at-music-festivals-are-threat-to-investors-as-well-as-fans.html?_r=0.

"Skream Talks Dubstep Warz, Hatcha and FWD." Red Bull Academy, September 24, 2011. https://youtu.be/wqchdQQV_QY.

"Soundgoodizer a Blessing or a Curse?" *The FLipside*, February 22, 2009. http://www.theflipsideforum.com/index.php?topic=11536.0.

"Sowall: From Jazz Chops to Finger Drumming." Ableton, October 1, 2019. https://www.ableton.com/en/blog/sowall-jazz-chops-finger-drumming/.

Spitz, Andrew. "Skube: A Last.fm & Spotify Radio." September 12, 2012. https://vimeo.com/50435491.

Stanyek, Jason, and Benjamin Piekut. "Deadness: Technologies of the Intermundane." *Drama Review* 54, no. 1 (Spring 2010): 14–38.

Stepulevage, L. "Gender/Technology Relations: Complicating the Gender Binary." *Gender and Education* 13, no. 3 (2001): 325–338.

Stern, Craig. "What Makes a Game Indie: A Universal Definition." *Sinister Design*, August 22, 2012. http://sinisterdesign.net/what-makes-a-game-indie-a-universal-definition/.

* Sterne, Jonathan. *The Audible Past: Cultural Origins of Sound Reproduction*. Durham, NC: Duke University Press, 2003.

Sterne, Jonathan. "Sonic Imaginations." In *The Sound Studies Reader*, edited by Jonathan Sterne, 1–17. New York: Routledge, 2012.

"Steve Jobs Introduces iMovie & iMac DV: Apple Special Event (1999)." EverySteveJobsVideo, December 21, 2012. https://www.dailymotion.com/video/x3m5log.

"Steve Jobs Introduces Rainbow iMacs & Power Mac G3: Macworld SF (1999)." EverySteveJobsVideo, uploaded December 21, 2013. https://www.dailymotion.com/video/x3mssqj.

Stewart, Rob. "AI-Based Mastering, Human Audio Mastering and the Future." *JustMastering*, May 29, 2016. https://www.justmastering.com/article-landr-audio-mastering.php.

Stockwell, Stephen, and Adam Muir. "The Military-Entertainment Complex: A New Facet of Information Warfare." *Fibreculture Journal* 1 (2003). http://one.fibreculture journal.org/fcj-004-the-military-entertainment-complex-a-new-facet-of-informat ion-warfare.

Stone, Allucquère Rosanne. *The War of Desire and Technology at the Close of the Mechanical Age*. Cambridge, MA: MIT Press, 1995.

Strachan, Robert. *Sonic Technologies: Popular Music, Digital Culture, and the Creative Process*. London: Bloomsbury, 2017.

Suits, Bernard. *The Grasshopper: Games, Life, and Utopia*. Peterborough, ON: Broadview, 2005.

"Summer Seasonal Gaming Guide." *Polygon*, n.d. https://www.polygon.com/2020/6/1/21273160/summer-seasonal-gaming-guide.

Supper, Alexandra. "The Search for the 'Killer Application': Drawing the Boundaries around the Sonification of Scientific Data." In *The Oxford Handbook of Sound Studies*, edited by Trevor Pinch and Karin Bijsterveld, 249–270. New York: Oxford University Press, 2012.

"Suzi Analogue: Staying in the Zone." Ableton, March 8, 2018. https://www.ableton.com/en/blog/suzi-analogue-staying-zone/.

Sweetser, Penny. *Emergence in Games*. Boston: Course Technology, 2008.

Tanaka, Atau. "Mapping Out Instruments, Affordances, and Mobiles." In *Proceedings of the 2010 Conference on New Interfaces for Musical Expression*, 15–18. Sydney: NIME, 2010.

Tanaka, Atau, Alessandro Altavilla, and Neal Spowage. "Gestural Musical Affordances." In *Proceedings of the 9th Sound and Music Computing Conference*, 318–325. Copenhagen: SMC, 2012.

Tarakajian, Sam. "An Interview with Ducky." Cycling '74, November 18, 2013. https://cycling74.com/2013/11/18/an-interview-with-ducky/.

Taylor, Robert W. "Hyper-Compression in Music Production: Agency, Structure and the Myth That 'Louder Is Better.'" *Journal on the Art of Record Production* 11 (2017). n.p.

https://www.arpjournal.com/asarpwp/hyper-compression-in-music-production-agency-structure-and-the-myth-that-louder-is-better/.

Taylor, Timothy. "The Commodification of Music at the Dawn of the Era of 'Mechanical Music." *Ethnomusicology* 51, no. 2 (2007): 281–305.

Taylor, Timothy. *The Sounds of Capitalism: Advertising, Music, and the Conquest of Culture*. Chicago: University of Chicago Press, 2012.

Taylor, Timothy, Mark Katz, and Tony Grajeda, eds. *Music, Sound, and Technology in America: A Documentary History of Early Phonograph, Cinema, and Radio*. Durham, NC: Duke University Press, 2012.

Terranova, Tiziana. "Free Labor: Producing Culture for the Digital Economy. *Social Text* 18, no. 2 (2000): 33–58.

Terranova, Tiziana. *Network Culture: Politics for the Information Age*. London: Pluto, 2004.

Théberge, Paul. *Any Sound You Can Imagine: Making Music/Consuming Technology*. Hanover, NH: Wesleyan University Press, 1997.

Théberge, Paul. "The Network Studio: Historical and Technological Paths to a New Ideal in Music Making." *Social Studies of Science* 34, no. 5 (2004): 759–781.

"35,000 Visitors at the Paris Maker Faire, a Record." *Makery*, May 4, 2015. http://www.makery.info/en/2015/05/04/35000-visiteurs-a-la-maker-faire-paris-un-record/.

Thompson, Clive. *Coders: The Making of a New Tribe and the Remaking of the World*. London: Penguin, 2019.

Thornton, Sarah. *Club Cultures: Music, Media, and Subcultural Capital*. Hanover, NH: Wesleyan University Press, 1996.

Tidwell, Jenifer. *Designing Interfaces: Patterns for Effective Interaction Design*. Sebastopol, CA: O'Reilly Media, 2005.

Tsioulcas, Anastasia. "A Year after the #MeToo Grammys, Women Are Still Missing in Music." NPR, February 8, 2019. https://www.npr.org/2019/02/08/692671099/a-year-after-the-metoo-grammys-women-are-still-missing-in-music.

Tucker, Boima. "Global Genre Accumulation." *Africa Is a Country*, November 22, 2011. https://africasacountry.com/2011/11/global-genre-accumulation/.

Tufte, Edward. "iPhone Resolution." November 25, 2008. https://www.youtube.com/watch?v=YslQ2625TR4.

Turing, Alan. "Computing Machinery and Intelligence." *Mind* 59, no. 236 (1950): 433–460.

Turkle, Sherry. *Alone Together: Why We Expect More from Technology and Less from Each Other*. New York: Basic Books, 2011.

Turkle, Sherry. "Computational Reticence: Why Women Fear the Intimate Machine." In *Technology and Women's Voices*, edited by Cheris Kramarae, 41–61. New York: Pergamon, 1986.

Turkle, Sherry. *The Second Self: Computers and the Human Spirit*. New York: Simon & Schuster, 1984.

Turner, Amber Leigh. "The History of Flat Design: How Efficiency and Minimalism Turned the Digital World Flat." *The Next Web*, March 19, 2014. http://thenextweb.com/dd/2014/03/19/history-flat-design-efficiency-minimalism-made-digital-world-flat/.

"25 Products That Changed Recording." *Sound on Sound*, November 2010. http://www.soundonsound.com/sos/nov10/articles/25-milestone-products.htm.

Unger, Russ, and Carolyn Chandler. *A Project Guide to UX Design: For User Experience Designers in the Field or in the Making*. Upper Saddle River, NJ: Pearson Education, 2009.

Upton, Brian. *The Aesthetic of Play*. Cambridge, MA: MIT Press, 2015.

Upbin, Bruce. "Why Los Angeles Is Emerging as the Next Silicon Valley." *Forbes*, August 28, 2012. http://www.forbes.com/sites/ciocentral/2012/08/28/why-los-angeles-is-emerging-as-the-next-silicon-valley/.

U.S. Senate. "Record Labeling: Hearing before the Committee on Commerce, Science, and Transportation." Ninety-Ninth Congress, First Session on Contents of Music and the Lyrics of Records, September 19, 1985. http://www.joesapt.net/superlink/shrg99-529/.

Vágnerová, Lucie. "'Nimble Fingers' in Electronic Music: Rethinking Sound through Neo-colonial Labour." *Organised Sound* 22, no. 2 (2018): 250–258.

Vale, V., and Andrea Juno, eds. Incredibly Strange Music, Vol. 1. San Francisco: RE/Search, 1993.

Van Den Boomen, Marianne. *Transcoding the Digital: How Metaphors Matter in New Media*. Amsterdam: Institute of Network Cultures, 2014.

Van Veen, Tobias, and Bernardo Alexander Attias. "Off the Record: Turntablism and Controllerism in the 21st Century (Part 1)." *Dancecult* 3, no. 1 (2011).

Veal, Michael. *Dub: Soundscapes and Shattered Songs in Jamaican Reggae*. Middletown, CT: Wesleyan University Press.

Vercellone, Carlo. "Wages and Rent: The New Articulation of Wages, Rent and Profit in Cognitive Capitalism." Generation Online, n.d. https://www.generation-online.org/c/fc_rent2.htm.

Vonderau, Patrick. "The Politics of Content Aggregation." *Television & New Media* 16, no. 8 (2015): 717–733.

Von Hornbostel, Erich M., and Curt Sachs. "Classification of Musical Instruments: Translated from the Original German by Anthony Baines and Klaus P. Wachsmann." *Galpin Society Journal* 14 (1961): 3–29.

Vorback, Kyle. "Ben Shapiro Produces EDM EP Using Logic and Reason." *The Hard Times*, July 12, 2019. https://thehardtimes.net/culture/ben-shapiro-produces-edm-ep-using-logic-and-reason/.

Wajcman, Judy. *Feminism Confronts Technology*. Cambridge: Polity, 1991.

Waksman, Steve. "Reading the Instrument: An Introduction." *Popular Music and Society* 26, no. 3 (2003): 251–261.

Wallace, Anne. "'Animal Crossing: New Horizons' Is Exactly What We Need Right Now." *Deseret News*, March 23, 2020. https://www.deseret.com/entertainment/2020/3/23/21191101/animal-crossing-new-horizons-nintendo-coronavirus-social-distancing-quarantine-covid-19.

Walther-Hansen, Mads. "New and Old User Interface Metaphors in Music Production." *Journal on the Art of Record Production* 11 (2017). n.p. https://www.arpjournal.com/asarpwp/new-and-old-user-interface-metaphors-in-music-production/.

Wang, Ge. "Ocarina: Designing the iPhone's Magic Flute." *Computer Music Journal* 38, no. 2 (2014): 8–21.

Wardrip-Fruin, Noah. *Expressive Processing: Digital Fictions, Computer Games, and Software Studies*. Cambridge, MA: MIT Press, 2009.

Watson, Sara M. "Data Doppelgängers and the Uncanny Valley of Personalization." *The Atlantic*, June 16, 2014. https://www.theatlantic.com/technology/archive/2014/06/data-doppelgangers-and-the-uncanny-valley-of-personalization/372780/.

Watts, Alan. *Nature, Man, and Woman*. New York: Vintage Books, 1991.

Weheliye, Alexander G. "Desiring Machines in Black Popular Music." In *The Sound Studies Reader*, edited by Jonathan Sterne, 511–519. New York: Routledge, 2012.

Weheliye, Alexander G. *Phonographies: Grooves in Sonic Afro-Modernity*. Durham, NC: Duke University Press, 2005.

Weheliye, Alexander G. "Rhythms of Relation: Black Popular Music and Mobile Technologies." *Current Musicology* 99–100 (2017): 107–127.

Weiser, Mark. "The Computer for the 21st Century." *Scientific American* 265, no. 3 (1991): 94–105.

Weiss, Dan. "The Unlikely Rise of FL Studio, the Internet's Favorite Production Software." *Vice*, October 12, 2016. https://noisey.vice.com/en_us/article/d33xzk/fl-studio-souljaboy-porter-robinson-madeon-feature.

Weiss, Jeff. "Flying Lotus' Nocturnal Visions." *LA Weekly*, October 4, 2012. http://www.laweekly.com/music/flying-lotus-nocturnal-visions-2611706.

"When Will the Bass Drop?" *Saturday Night Live*, NBC, May 18, 2014. https://youtu.be/DoUV7Q1C1SU.

Whitelaw, Mitchell. "System Stories and Model Worlds: A Critical Approach to Generative Art." In *Readme 100: Temporary Software Art Factory*, edited by Olga Goriounova, 135–154. Norderstedt: BoD, 2005.

Whiteley, Sheila. *Women and Popular Music: Sexuality, Identity, and Subjectivity*. London: Psychology Press, 2000.

"Why Do We Exist?" CloudBounce, n.d. https://www.cloudbounce.com/about#why.

Wiener, Norbert. "Men, Machines, and the World About." In *Medicine and Science*, edited by I. Galderston, 13–28. New York: International Universities Press, 1954.

Wilken, Rowan. "From *Stabilitas Loci* to *Mobilitas Loci*: Networked Mobility and the Transformation of Place." *Fibreculture Journal* 6 (2005). https://six.fibreculturejournal.org/fcj-036-from-stabilitas-loci-to-mobilitas-loci-networked-mobility-and-the-transformation-of-place/.

Williams, Alan. "Putting It on Display: The Impact of Visual Information on Control Room Dynamics." *Journal on the Art of Record Production* 6 (2012): n.p. https://www.arpjournal.com/asarpwp/putting-it-on-display-the-impact-of-visual-information-on-control-room-dynamics/.

Williams, Raymond. *Television, Technology and Cultural Form*, 3rd ed. New York: Routledge, 2003.

Wilson, Scott. "The 14 Drum Machines That Shaped Modern Music." *FACTmag*, September 22, 2016. https://www.factmag.com/2016/09/22/the-14-drum-machines-that-shaped-modern-music/.

Wing, Jeanette. "Computational Thinking." *Communications of the ACM* 49, no. 3 (2006): 33–35.

Winters, Patrick. *Sound Design for Low and No Budget Films*. New York: Routledge, 2017.

Wodtke, Christina. *Information Architecture: Blueprints for the Web*. San Francisco: New Riders, 2009.

Wojcicki, Susan. "Closing the Tech Industry Gender Gap." *Huffington Post*, January 27, 2016. http://www.huffingtonpost.com/susan-wojcicki/tech-industry-gender-gap_b_9089472.html.

Wolf, Zachary B. "This Is What Coronavirus Capitalism Looks Like." CNN, April 28, 2020. https://www.cnn.com/2020/04/28/politics/what-matters-april-27/index.html.

Wolfe, Paula. "A Studio of One's Own: Music Production, Technology, and Gender." *Journal on the Art of Record Production* 7 (2012): n.p. https://www.arpjournal.com/asarpwp/a-studio-of-ones-own-music-production-technology-and-gender/.

Wooller, Rene, Andrew R. Brown, Eduardo Miranda, Rodney Berry, and Joachim Diederich. "A Framework for Comparison of Processes in Algorithmic Music Systems." In *Proceedings from Generative Arts Practice*, 109–124. Sydney: Creativity and Cognition Studios Press, 2005.

Wozniak, Steve. *iWoz*. New York: W. W. Norton, 2006.

Wurman, Richard Saul. *Information Architects*. New York: Graphis, 1997.

Young, John. "Sound Morphology and the Articulation of Structure in Electroacoustic Music." *Organised Sound* 9, no. 1 (2004): 7–14.

Zak, Albin J. *The Poetics of Rock: Cutting Tracks, Making Records*. Berkeley: University of California Press, 2001.

Zicarelli, David. "How I Learned to Love a Program That Does Nothing." *Computer Music Journal* 26, no. 4 (2002): 44–51.

Zicarelli, David. "An Interview with William Kleinsasser." Cycling '74, September 13, 2005. https://cycling74.com/2005/09/13/an-interview-with-william-kleinsasser/.

Zittrain, Jonathan. *The Future of the Internet—And How to Stop It*. New Haven, CT: Yale University Press, 2008.

Index

For the benefit of digital users, indexed terms that span two pages (e.g., 52–53) may, on occasion, appear on only one of those pages.

Figures and tables are indicated by *f* and *t* following the page number.

CPSIA information can be obtained
at www.ICGtesting.com
Printed in the USA
BVHW030748261022
650288BV00001B/1